Springer Theses

Recognizing Outstanding Ph.D. Research

Aims and Scope

The series "Springer Theses" brings together a selection of the very best Ph.D. theses from around the world and across the physical sciences. Nominated and endorsed by two recognized specialists, each published volume has been selected for its scientific excellence and the high impact of its contents for the pertinent field of research. For greater accessibility to non-specialists, the published versions include an extended introduction, as well as a foreword by the student's supervisor explaining the special relevance of the work for the field. As a whole, the series will provide a valuable resource both for newcomers to the research fields described, and for other scientists seeking detailed background information on special questions. Finally, it provides an accredited documentation of the valuable contributions made by today's younger generation of scientists.

Theses may be nominated for publication in this series by heads of department at internationally leading universities or institutes and should fulfill all of the following criteria

- They must be written in good English.
- The topic should fall within the confines of Chemistry, Physics, Earth Sciences, Engineering and related interdisciplinary fields such as Materials, Nanoscience, Chemical Engineering, Complex Systems and Biophysics.
- The work reported in the thesis must represent a significant scientific advance.
- If the thesis includes previously published material, permission to reproduce this must be gained from the respective copyright holder (a maximum 30% of the thesis should be a verbatim reproduction from the author's previous publications).
- They must have been examined and passed during the 12 months prior to nomination.
- Each thesis should include a foreword by the supervisor outlining the significance of its content.
- The theses should have a clearly defined structure including an introduction accessible to new PhD students and scientists not expert in the relevant field.

Indexed by zbMATH.

More information about this series at http://link.springer.com/bookseries/8790

W. Wicak Ananduta

Non-centralized Optimization-Based Control Schemes for Large-Scale Energy Systems

Doctoral Thesis accepted by
Universitat Politècnica de Catalunya, Barcelona, Spain

Springer

Author
Dr. W. Wicak Ananduta
Delft, The Netherlands

Department of Automatic Control
Universitat Politècnica de Catalunya (UPC)
Barcelona, Spain

Supervisor
Prof. Carlos Ocampo-Martinez
Department of Automatic Control
Universitat Politècnica de Catalunya (UPC)
Barcelona, Spain

ISSN 2190-5053 ISSN 2190-5061 (electronic)
Springer Theses
ISBN 978-3-030-89802-1 ISBN 978-3-030-89803-8 (eBook)
https://doi.org/10.1007/978-3-030-89803-8

This Springer imprint is published by the registered company Springer Nature Switzerland AG
The registered company address is: Gewerbestrasse 11, 6330 Cham, Switzerland

To Desy

To Tara

To my parents

Supervisor's Foreword

Scientific research is an adventure. Sometimes, adventures are risky, boring, exciting or just life experiences. However, when people love what they do, nice adventures are part of the way they walk everyday. In research, loving the topic one is focused on making everything quite funny and exciting. This sensation in the stomach pushes forward other important feelings, e.g. curiosity, expectation and nonconformity. And then, the proper mixtures of those sensations/feelings give raise to the potential success. However, as a part of this kind of adventures, the right exploring is a quite important pillar when looking for answers.

When this thesis started, there were two explorers: the former with some ideas in his mind, with some queries with no answer yet, with passion for his research topics. The latter, but not less important, seemed to be someone who wanted to get over the world with happiness and curiosity, with devotion and eager to learn a lot, with all these sensations and feelings needed to reach success. It was the genesis of this brilliant research work.

At the beginning, this thesis was conceived to be focused on management of large-scale flow-based networks, being the energy systems the particular area of interest. Tons of research works and papers have reported interesting results in this area. However, there were two main aspects to consider: reliability and, intrinsically, modularity. These considered aspects implicitly traced the way of developing all the research done. Hence, two stages can be described. The first stage collected all the results related to resilient methods in electrical networks, which look for the robustification of the decisions made against adversarial behaviours through performing an identification/mitigation method based on hypothesis testing using Bayesian inference. Interesting and relevant results were reached in a joint work with Profs. Hideaki Ishii (Japan) and José Maestre (Spain).

The second stage explored around the modularity concept. As it is well known, several research works cope with the non-centralised control problem when proposing solutions for large-scale network systems. However, few of them treat the quite relevant problem of how to split the system into subsystems and hence designing non-centralised controllers. Therefore, this stage of the thesis started with the idea of determining partitioning strategies that, together with a suitable optimization-based control design, would propose a proper management strategy for electric networks considering diverse control objectives. To this end, the problem was attacked since its roots by studying and proposing relevant results on distributed optimization towards stating and solving problems related to partitioning and predictive control of such systems. Most of this stage was covered in a joint work with Prof. Angelia Nedić (USA).

In the end, these two stages converged into this document, which gives the clear and mathematically formal statements for the analysis and design of non-centralised predictive controllers applied to flow-based energy networks taking into account resiliency capabilities. The controllers are proposed and designed based on the fact that the large-scale system is properly split into subsystems that make such a controller design simpler and modular. All the results and proposed approach derived from this complete thesis open a very interesting way for the deeper analysis and formalisation of partitioning approaches towards their proper synergy with optimization-based controller designs towards a new generation of control strategies supported by distributed optimization.

This adventure ended up successfully. The explorers learnt a lot from each other. The curiosity inspired very nice discussions and drove the explorers to visit distant lands bringing back new ideas and motivations. Finally, enjoy what you do is the first step that makes the others appreciate your effort.

Barcelona, Spain
August 2020

Prof. Carlos Ocampo-Martinez

Parts of this thesis have been published in the following articles:

Journals

- W. Ananduta, J. Barreiro-Gomez, C. Ocampo-Martinez, and N. Quijano. Mitigation of communication failures in distributed model predictive control strategies. *IET Control Theory Applications*, 12(18):2507–2515, 2018.
- W. Ananduta, T. Pippia, C. Ocampo-Martinez, J. Sijs, and B. De Schutter. Online partitioning method for decentralized control of linear switching large-scale systems. *Journal of the Franklin Institute*, 356(6):3290–3313, 2019.
- W. Ananduta, J. M. Maestre, C. Ocampo-Martinez, and H. Ishii. Resilient distributed model predictive control for energy management of interconnected microgrids. *Optimal Control, Applications and Methods*, 41(1):146–169, 2020.
- W. Ananduta, C. Ocampo-Martinez, and A. Nedić. A distributed augmented Lagrangian method over stochastic network for economic dispatch of large-scale energy systems. *IEEE Transactions on Sustainable Energy*, 2021. in press. Available online at https://doi.org/10.1109/TSTE.2021.3073510.
- W. Ananduta, A. Nedić, and C. Ocampo-Martinez. Distributed augmented Lagrangian method for link-based resource sharing problems of multi-agent systems. *IEEE Transactions on Automatic Control*. in press. Available online at https://doi.org/10.1109/TAC.2021.3092561.
- W. Ananduta and C. Ocampo-Martinez. Event-triggered partitioning for non-centralized predictive-control-based economic dispatch of interconnected microgrids. *Automatica*, 2021. in press. Available online at https://doi.org/10.1016/j.automatica.2021.109829.

Conferences

- W. Ananduta, J. Barreiro-Gomez, C. Ocampo-Martinez, and N. Quijano. Resilient information-exchange protocol for distributed model predictive control schemes. In *Proceedings of the American Control Conference (ACC)*, pp. 1286–1291, 2018.
- W. Ananduta and C. Ocampo-Martinez. Decentralized energy management of power networks with distributed generation using periodical self-sufficient repartitioning approach. In *Proceedings of the American Control Conference (ACC)*, pp. 3230–3235, 2019.
- W. Ananduta, J. M. Maestre, C. Ocampo-Martinez, and H. Ishii. Resilient distributed energy management for systems of interconnected microgrids. In *Proceedings of the IEEE 57th Conference on Decision and Control (CDC)*, pp. 6159–6164, Miami, USA, 2018.
- W. Ananduta, J. M. Maestre, C. Ocampo-Martinez, and H. Ishii. A resilient approach for distributed MPC-based economic dispatch in interconnected microgrids. In *Proceedings of the European Control Conference*, pp. 891–896, Naples, Italy, 2019.

Acknowledgements

I want to express my sincerest gratitude to those who are involved in my endeavour to finish this manuscript.

Firstly, my gratitude is to my advisor, Prof. Carlos Ocampo-Martinez. His guidance and total support have made this thesis possible. He has always been available every single time I seek for advice, which is not only about the thesis but also life in general. He has given me a lot of opportunities and feedback to grow. For that, I am forever indebted.

Secondly, I thank Prof. Angelia Nedić for her patience during our discussions and invaluable experience at Arizona State University. I also thank Prof. Hideaki Ishii and Prof. José Maria Maestre, who received me as a visiting researcher at the Tokyo Institute of Technology. Their ideas, expertise and insights have helped in shaping my thesis.

I also thank the innovative controls for renewable source integration into smart energy systems (INCITE) training network, which was funded by the European Union through Marie Skłodowska-Currie grant agreement No. 675318, for providing financial support, trainings and workshops. More importantly, through this network, I have met other INCITE fellows: Tomás, Unni, Jesus, Sara, Thibault, Ade, Shantanu, Hazem, Nikolaos, Felix, Kostas, Camilo, Paschalia, and Miguel, as well as with INCITE's manager, Marta. I really enjoyed the activities that we have spent together.

Moreover, I am grateful that I have had Dr. Julián Barreiro-Gomez, Dr. Masoud Pourasghar, Dr. Pablo Segovia, Dr. Ye Wang, Jenny Diaz and Fatemeh Karimi as my officemates, with whom I have shared my Ph.D. journey.

Finally, I owe my gratitude to my family. Thanks to my parents, for their infinite love and everything that they have done, and to my brothers, for their support. I also dedicate this thesis to my late grandmother, who was crucial to my upbringing. Most importantly, to my wife, Desy, and our greatest joy, Tara, no word can describe how significant you are to me. I would not have survived this journey without you.

Barcelona, Spain
November 2019

W. Wicak Ananduta

Contents

Notations

$\mathbb{1}_n$	All-one column vector with the size of n
I_n	Identity matrix with the size of $n \times n$
$\mathbb{E}(\cdot)$	Expectation operator
$\mathbb{P}(\cdot)$	Probability measure
$\mathbb{R}$	Set of real numbers
$\mathbb{R}_{\geq a}$	$\{b \in \mathbb{R} : b \geq a\}$, for $a \in \mathbb{R}$
$\mathbb{R}_{>a}$	$\{b \in \mathbb{R} : b > a\}$, for $a \in \mathbb{R}$
$\mathbb{Z}$	Set of integers
$\mathbb{Z}_{\geq a}$	$\{b \in \mathbb{Z} : b \geq a\}$, for $a \in \mathbb{Z}$
$\mathbb{Z}_{>a}$	$\{b \in \mathbb{Z} : b > a\}$, for $a \in \mathbb{Z}$
$\prod$	Cartesian product
$A \otimes B$	Kronecker product of two matrices A and B
$\langle a, b \rangle$	Inner product of $a, b \in \mathbb{R}^n$
$\lvert \cdot \rvert$	Cardinality operator
$\lVert \cdot \rVert_2$	Euclidean norm
$\lVert \cdot \rVert_D^2$	The squared of weighted norm induced by the square matrix D, e.g. for $a \in \mathbb{R}^n$ and $D \in \mathbb{R}^{n \times n}$, $\lVert a \rVert_D^2 = \langle a, Da \rangle$.
blkdiag$(\cdot)$	Construct a block diagonal matrix of the arguments
col$(\cdot)$	Concatenate the arguments as a column vector
diag$(\cdot)$	Construct a diagonal matrix of the arguments

Chapter 1
Introduction

This chapter provides the background and motivation of this thesis by illustrating the development of the energy sector, stating some new control challenges that arise due to the transformation of energy systems, and introducing the model predictive control (MPC) approach that is suitable as an energy management framework and able to overcome the challenges. Moreover, the key research questions, which incite the development of the thesis, are formulated. Finally, the contents and the related publications of the thesis are stated.

1.1 Development of Energy Systems

Current innovations in the energy sector are driven to achieve clean, efficient, and sustainable production, operation, and consumption of energy, as targeted by policy-makers from many countries, e.g., as stated in [1–5]. Significant changes in energy systems that can be seen nowadays include the penetration of small-scale production units, particularly those that use renewable energy sources, at a high rate [6], the introduction of flexible (controllable) loads, not only in households but also in the form of electrical vehicles [7], and the integration with information and communication infrastructures to set up intelligent systems [8].

A paradigm shift on the way energy systems must be controlled and managed is required to take full advantage of the above technological developments. For instance, the power generated by non-dispatchable renewable energy sources, such as solar and wind, is intermittent and weather-dependent unlike conventional fossil-based power generators; thus, they require storage devices to compensate for this limitation [9]. Moreover, the hierarchical and centralized control approach in electrical transmission and distribution networks might not be applicable and must be adapted when a large number of distributed generation and storage units as well as flexible loads

W. W. Ananduta, *Non-centralized Optimization-Based Control Schemes for Large-Scale Energy Systems*, Springer Theses,
https://doi.org/10.1007/978-3-030-89803-8_1

are present in the system [10, 11]. Additionally, the availability of information and communication technology encourages the development of novel approaches for controlling and managing energy systems [9, 10].

The emphasis of this thesis is on the economic dispatch problem of energy networks considering the aforementioned technological development. Specifically, the regarded energy networks may consist of a large number of controllable components, such as distributed generation units, which can be dispatchable or non-dispatchable, storage devices, and flexible loads. The economic dispatch problem of such systems deals with computing economically optimal references of the controllable components such that the generated power meets the demand and the operational constraints of the network are satisfied [12]. Therefore, the problem can be considered as an optimization problem of a large-scale system.

One of the main challenges faced when dealing with a control problem of a large-scale system is the complexity of computing control inputs, especially when the available computational time and resources are limited. Moreover, a large amount of data, such as data from sensors or control inputs for the actuators, must also be communicated between the system and the controller. Additionally, for some large-scale systems, including energy networks, the reliability, the scalability, and the flexibility of the controller are also important [13]. Furthermore, such systems may also have additional features, such as uncertain behavior of their components and time-varying topologies, which increase the complexity of the control problem. In an energy network, not only the energy demand but also the energy produced by non-dispatchable generation units, such as solar- or wind-powered units, are uncertain. Moreover, current electrical grids may also possess time-varying topologies. For instance, electric vehicles, as ones of the components of the grid, move around and are connected to different charging points at different time instants.

1.2 Non-centralized Model Predictive Control

Model Predictive Control (MPC) or receding horizon control is an online optimization-based control method that computes its control inputs by minimizing a cost function while taking into account the dynamics of the system and both physical and operational constraints [14]. The controller requires the model of the systems in order to represent its dynamical behavior. Furthermore, an MPC controller takes into account the prediction over a certain time horizon and only applies the control input that corresponds to the current time instant. Additionally, it also applies the receding horizon principle, in which the prediction horizon is always shifted forward at each time instant. The MPC framework, with its feature, is suitable to be implemented as an economic dispatch scheme of energy networks. One of the advantages of considering the MPC framework is in handling components with dynamics and uncertainties, in comparison to tradional economic dispatch schemes [15, 16]. In this regard, the energy management problem must be cast as a finite horizon optimization problem, which includes the economic cost function as well as the physical and the operational constraints of the components in the system.

In dealing with a large network, the MPC framework has been developed to incorporate non-centralized architectures, which are defined as follows:

Definition 1.1 (*Non-centralized architecture*) A control scheme has a non-centralized architecture if there exists a group of local controllers, each of which is responsible for a partition/portion of the system. □

Note that each local controller may only require partial model, constraints, and states as well as compute partial control inputs. Furthermore, following [13, 17], non-centralized MPC schemes are classified into the decentralized and distributed architectures, which are defined next.

Definition 1.2 (*Decentralized architecture* [13]) A non-centralized control architecture is called decentralized if the local controllers do not communicate with each other to compute their control inputs. □

Definition 1.3 (*Distributed architecture* [13]) A non-centralized control architecture is called distributed if the local controllers communicate with each other to compute their control inputs. □

Decentralized and distributed schemes are shown in Figs. 1.1 and 1.2, respectively. As provided in their definitions, the main distinction of these two architectures is the existence of communication when computing the control inputs. Note that communication is an important feature of distributed approaches since it allows local controllers to have extra information about their neighbors. Therefore, the performance of distributed approaches is, in general, better than the decentralized counterparts. In fact, most of the non-centralized MPC approaches that are able to meet the centralized performance belong to the class of distributed approaches. However, decentralized controllers are relatively more straightforward than distributed ones while achieving a suboptimal performance that may be enough for specific systems, e.g., weakly coupled systems [13].

Non-centralized schemes are developed mainly to overcome issues faced by the centralized counterpart when the system is too large and complex [13, 17]. One of

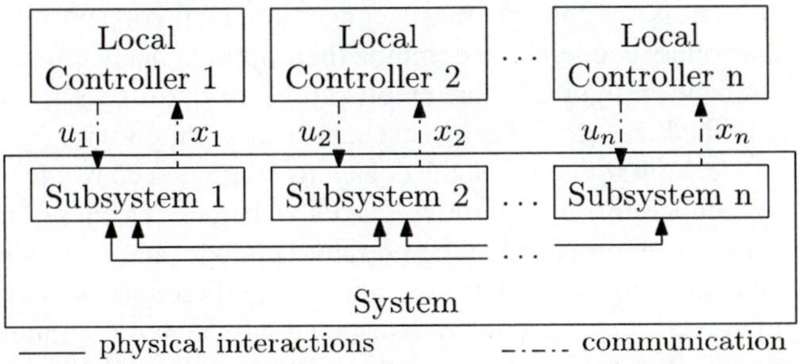

Fig. 1.1 Decentralized control scheme

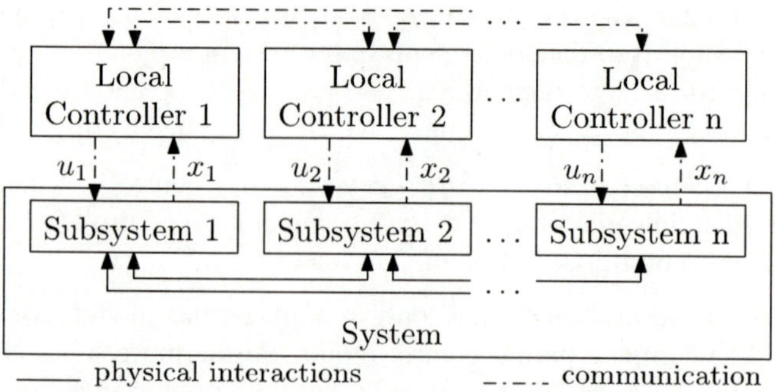

Fig. 1.2 Distributed control scheme

the problems is intractability, which is the inability of the controller to compute the control input in a given time. This issue often arises since the optimization problem behind the controller is too large or too complex. In this regard, by having a non-centralized scheme, the optimization problem can be decomposed into subproblems, which are locally tractable and then assigned to the local controllers. Furthermore, as discussed in [13, 17], non-centralized schemes are more flexible and scalable than the centralized counterpart.

Summaries and surveys of non-centralized MPC approaches are provided in [13, 17–19]. Furthermore, in [18, 19], some classifications are also made. These approaches are classified based on the process features, such as coupling source; control features, such as architecture, controller knowledge and attitude, computation type, or communication; and theoretical features, such as optimality and robustness. In designing a non-centralized scheme, it is important to identify the coupling sources, i.e., what makes the overall system non-separable. Based on this knowledge and the control objectives, then some control features can be decided. For instance: whether the information required by each controller is strictly local, i.e., only the information of the neighboring subsystems, or partially global, i.e., some information of all subsystems; whether the attitude is cooperative or non-cooperative; and whether the communication between the controllers is serial or parallel.

Based on the nature of the considered control problem, where an optimal decision of a global problem is sought, this thesis concentrates on cooperative approaches, where local controllers cooperate to compute their optimal decisions. In this regard, distributed optimization algorithms are employed. Such algorithms are typically iterative and require the local controllers to exchange information with their neighbors. Furthermore, under some conditions, the decisions computed converge to an optimal solution. Some cooperative and iterative DMPC schemes for energy management problems that have been proposed in the literature include, among others, those that are based on dual decomposition [20], the alternating direction method of multipliers (ADMM) [21, 22], the accelerated distributed augmented Lagrangian (ADAL) [23], the optimality-condition decomposition (OCD) [24], and Douglas-Rachford operator splitting [25]. Furthermore, in [15], the optimization problem behind the

DMPC scheme is formulated as a mixed-integer programming problem, and the scheme uses a coordinator and consists of three optimization phases. Similarly, [26] also proposes a hierarchical distributed MPC scheme with a coordinator and two optimization phases.

Despite many efforts to employ a non-centralized MPC scheme in energy networks that have been made, further study and investigation are still needed. This thesis, in particular, gives special attention to the improvement of communication and cooperation processes that local controllers must carry out. As previously discussed, communication and cooperation play an important role in the deployment of a non-centralized scheme. This thesis addresses some potential communication and cooperation problems that might occur and develops methodologies to deal with them. Specifically, the problems include:

- Communication link failures, which prevent local controllers from obtaining necessary information.
- Asynchronous updates, where some local controllers are not able to send information to their neighbors.
- Non-compliance, where some local controllers do not implement the decisions that have been computed from applying a distributed optimization algorithm.

Furthermore, this thesis also exploits the system partitioning approach with the aim to reduce the intensity of the communication process when a distributed optimization algorithm is deployed.

1.3 Research Questions

The research objectives of this thesis are driven from the following questions:

(Q_1) How to design and implement a non-centralized MPC-based economic dispatch scheme for large-scale energy systems?
(Q_2) How to design an information-exchange protocol for distributed MPC approaches that is resilient against communication failures?
(Q_3) Do distributed optimization algorithms work over time-varying communication networks and in an asynchronous manner?
(Q_4) How to perform online partitioning on large-scale energy systems in a distributed manner?
(Q_5) How can online partitioning be used in a non-centralized economic dispatch scheme of large-scale energy systems?
(Q_6) How to design a non-centralized MPC-based economic dispatch scheme that is resilient against non-compliance issues?

While the *key research question* (Q_1) serves as the main objective of this thesis, the other questions provide specific goals. The remaining chapters of the thesis are devoted to providing the answers, which become the contributions of this thesis.

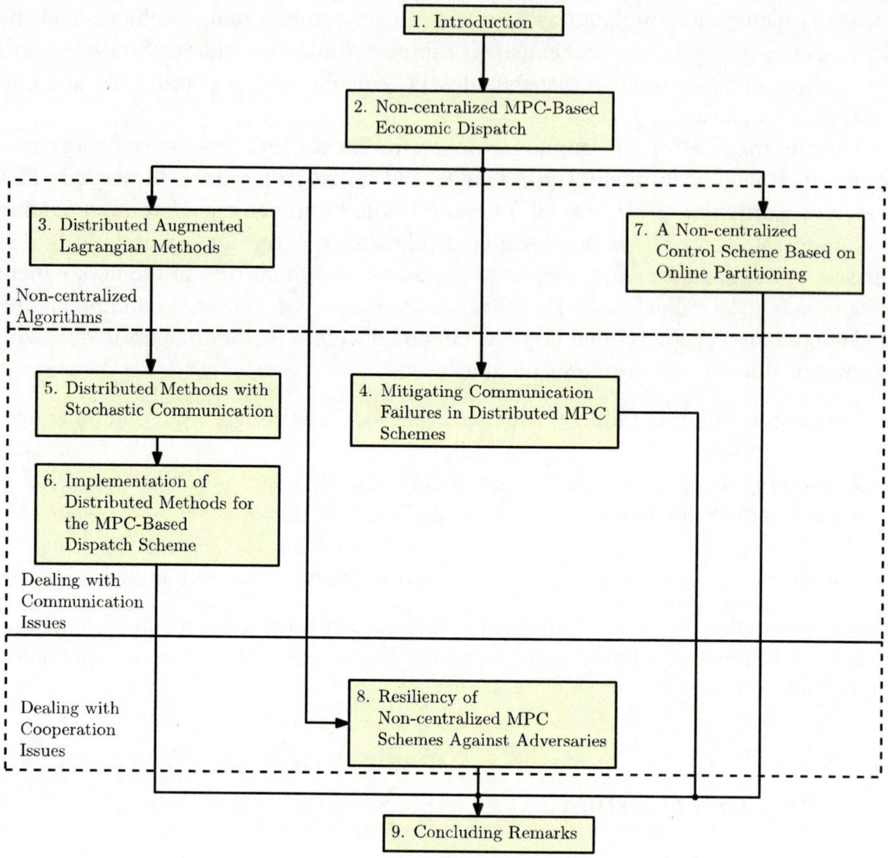

Fig. 1.3 Outline of the thesis. Arrows indicate *read-before* relations

1.4 Thesis Outline

Figure 1.3 presents the outline of the thesis and the connections of the chapters. Furthermore, the content of each chapter in this thesis is summarized as follows:

Chapter 2: Non-centralized MPC-based economic dispatch
This chapter provides the formulation of the energy management problem of a large-scale energy system. The model of the network, the optimization problem, and the general assumptions considered throughout the thesis are presented. Moreover, the non-centralized MPC framework and its specific issues that become key points in the discussion of this thesis are also explained. Therefore, this chapter initiates the answer to the *key research question* (Q_1).

Chapter 3: Distributed augmented Lagrangian methods
In this chapter, two distributed optimization methods, which are based on the augmented Lagrangian approach, are developed to solve the economic dispatch problem

formulated in Chap. 2. The distributed methods exploit the particular structure of the problem to achieve scalability. Furthermore, the convergence analysis of the proposed methods is also presented. This chapter partially answers *key research question* (Q_1) and is partly based on the following publication:

- W. Ananduta, A. Nedić, and C. Ocampo-Martinez. Distributed augmented Lagrangian method for link-based resource sharing problems of multi-agent systems. *IEEE Transactions on Automatic Control*. in press. Available online at https://doi.org/10.1109/TAC.2021.3092561.

Chapter 4: Mitigating communication failures in distributed MPC schemes
This chapter deals with the problem of communication failures that distributed MPC schemes might face in practice. Furthermore, it presents an application of consensus algorithm as the information-exchange protocol and a solution to the communication failure problem. Regarding the proposed solution, a discussion on how to partition the communication network and reconfigure the partitions is also provided. Thus, this chapter answers the *key research question* (Q_2). Moreover, it is based on the following publications:

- W. Ananduta, J. Barreiro-Gomez, C. Ocampo-Martinez, and N. Quijano. Resilient information-exchange protocol for distributed model predictive control schemes. In *Proceedings of the American Control Conference (ACC)*, pages 1286–1291, 2018.
- W. Ananduta, J. Barreiro-Gomez, C. Ocampo-Martinez, and N. Quijano. Mitigation of communication failures in distributed model predictive control strategies. *IET Control Theory Applications*, 12(18):2507–2515, 2018.

Chapter 5: Distributed methods with stochastic communication
This chapter investigates how to extend the distributed optimization methods presented in Chap. 3 such that they work under imperfect communication conditions. Specifically, the distributed methods are modified and implemented over a random communication network and in an asynchronous manner. Moreover, this chapter also studies their convergence properties. This chapter answers the *key research question* (Q_3) and is partially based on the following publication:

- W. Ananduta, A. Nedić, and C. Ocampo-Martinez. Distributed augmented Lagrangian method for link-based resource sharing problems of multi-agent systems. *IEEE Transactions on Automatic Control*. in press. Available online at https://doi.org/10.1109/TAC.2021.3092561.

Chapter 6: Implementation of distributed methods for the MPC-based dispatch scheme
This chapter studies the application of the distributed methods discussed in Chaps. 3 and 5 to the non-centralized MPC-based economic dispatch scheme for energy systems. Some techniques to improve the convergence speed, which is important for online optimization, are applied. The study is based on numerical simulations that are carried using a benchmark case. Therefore, this chapter partially answers

the *key research questions* (Q_1) and (Q_3). Additionally, it is based on the following publication:

- W. Ananduta, C. Ocampo-Martinez, and A. Nedić. A distributed augmented Lagrangian method over stochastic network for economic dispatch of large-scale energy systems. *IEEE Transactions on Sustainable Energy*, 2021. in press. Available online at https://doi.org/10.1109/TSTE.2021.3073510.

Chapter 7: A non-centralized control scheme based on online partitioning
This chapter discusses a distributed methodology to perform online system partitioning of large-scale energy networks. The partitioning method is specifically developed as a part of a non-centralized economic dispatch scheme, which has low communication burden. Furthermore, some analysis on the outcome of the partitioning method and the non-centralized control scheme is also provided. This chapter answers the *key research questions* (Q_1), (Q_4), and (Q_5). Furthermore, it is based on the following publications:

- W. Ananduta and C. Ocampo-Martinez. Decentralized energy management of power networks with distributed generation using periodical self-sufficient repartitioning approach. In *Proceedings of the American Control Conference (ACC)*, pages 3230–3235, 2019.
- W. Ananduta and C. Ocampo-Martinez. Event-triggered partitioning for non-centralized predictive-control-based economic dispatch of interconnected microgrids. *Automatica*, 2021. in press. Available online at https://doi.org/10.1016/j.automatica.2021.109829.

Chapter 8: Resiliency of non-centralized MPC schemes against adversaries
This chapter addresses cooperation issues that the non-centralized MPC-based economic dispatch scheme might encounter. Particularly, it is considered that some agents are adversarial and do not comply with the decisions computed by the non-centralized algorithm applied by the agents in order to have a better performance. Therefore, a method to cope with this issue is proposed. The method consists of robustifying the decisions and identifying as well as mitigating the adversarial behavior. This chapter answers the *key research question* (Q_6) and is based on the following publications:

- W. Ananduta, J. M. Maestre, C. Ocampo-Martinez, and H. Ishii. Resilient distributed energy management for systems of interconnected microgrids. In *Proceedings of the IEEE 57th Conference on Decision and Control (CDC)*, pages 6159–6164, Miami, USA, 2018.
- W. Ananduta, J. M. Maestre, C. Ocampo-Martinez, and H. Ishii. A resilient approach for distributed MPC-based economic dispatch in interconnected microgrids. In *Proceedings of the European Control Conference*, pages 891–896, Naples, Italy, 2019.
- W. Ananduta, J. M. Maestre, C. Ocampo-Martinez, and H. Ishii. Resilient distributed model predictive control for energy management of interconnected microgrids. *Optimal Control, Applications and Methods*, 41(1):146–169, 2020.

Chapter 9: Concluding remarks
This chapter provides some remarks regarding the results obtained and presented in Chaps. 2–8. The key research questions, introduced in Sect. 1.3, are also addressed. Furthermore, this chapter also suggests some open questions for future research.

1.5 Other Publication

The following publication is related to the research topic and has been done during the completion of this thesis:

- W. Ananduta, T. Pippia, C. Ocampo-Martinez, J. Sijs, and B. De Schutter. Online partitioning method for decentralized control of linear switching large-scale systems. *Journal of the Franklin Institute*, 356(6):3290–3313, 2019.

References

1. European Parliament and Council of European Union (2018) Directive EU no. 2018/2001. Off J Eur Union 82–209. https://eur-lex.europa.eu/legal-content/EN/TXT/PDF/?uri=CELEX:32018L2001&from=EN
2. European Parliament and Council of European Union (2018) Directive EU no. 2018/2002. Off J Eur Union 210–230. https://eur-lex.europa.eu/legal-content/EN/TXT/PDF/?uri=CELEX:32018L2002&from=EN
3. Department of the Environment and Energy of Australia. The renewable energy target (RET) scheme. https://www.environment.gov.au/climate-change/government/renewable-energy-target-scheme
4. National Conference of State Legislatures. State renewable portfolio standards and goals. http://www.ncsl.org/research/energy/renewable-portfolio-standards.aspx
5. Olivares DE, Mehrizi-Sani A, Etemadi AH, Cañizares CA, Iravani R, Kazerani M, Hajimiragha AH, Gomis-Bellmunt O, Saeedifard M, Palma-Behnke R, Jimènez-Estèvez GA, Hatziargyriou ND (2014) Trends in microgrid control. IEEE Trans Smart Grid 5(4):1905–1919
6. International Energy Agency (2019) Renewables 2018: market analysis and forecast from 2018 to 2023. https://www.iea.org/renewables2018/
7. Gorner M, Cazzola P, Scheffer S (2019) Electric vehicles: tracking clean energy progress. https://www.iea.org/tcep/transport/electricvehicles/
8. Amin SM (2011) Smart grid: overview, issues and opportunities. advances and challenges in sensing, modeling, simulation, optimization and control. Eur J Control 17(5–6):547–567
9. Morstyn T, Hredzak B, Agelidis VG (2018) Control strategies for microgrids with distributed energy storage systems: an overview. IEEE Trans Smart Grid 9(4):3652–3666
10. Molzahn DK, Dörfler F, Sandberg H, Low SH, Chakrabarti S, Baldick R, Lavaei J (2017) A survey of distributed optimization and control algorithms for electric power systems. IEEE Trans Smart Grid 8(6):2941–2962
11. Dörfler F, Simpson-Porco JW, Bullo F (2016) Breaking the hierarchy: distributed control and economic optimality in microgrids. IEEE Trans Control of Network Syst 3(3):241–253
12. Kar S, Hug G, Mohammadi J, Moura JMF (2014) Distributed state estimation and energy management in smart grids: a consensus+innovations approach. IEEE J Selected Topics Signal Process 8(6):1022–1038

13. Christofides PD, Scattolini R, Muñoz de la Peña D, Liu J (2013) Distributed model predictive control: a tutorial review and future research directions. Comput Chem Eng 51:21–41
14. Rawlings JB, Mayne DQ (2009) Model predictive control: theory and design. Nob Hill Publications, Madison, Wisconsin
15. Parisio A, Wiezorek C, Kyntäjä T, Elo J, Strunz K, Johansson KH (2017) Cooperative MPC-based energy management for networked microgrids. IEEE Trans Smart Grid 8(6):3066–3074
16. Zhu D, Hug G (2014) Decomposed stochastic model predictive control for optimal dispatch of storage and generation. IEEE Trans Smart Grid 5(4):2044–2053
17. Scattolini R (2009) Architectures for distributed and hierarchical model predictive control—a review. J Process Control 19(5):723–731
18. Maestre JM, Negenborn RR (eds) (2013) Distributed model predictive control made easy, vol 69. Springer Science & Business Media, Berlin
19. Negenborn RR, Maestre JM (2014) Distributed model predictive control: an overview and roadmap of future research opportunities. IEEE Control Syst Mag 34(4):87–97
20. Larsen GKH, van Foreest ND, Scherpen JMA (2014) Distributed MPC applied to a network of households with micro-CHP and heat storage. IEEE Trans Smart Grid 5(4):2106–2114
21. Wang T, O'Neill D, Kamath H (2015) Dynamic control and optimization of distributed energy resources in a microgrid. IEEE Trans Smart Grid 6(6):2884–2894
22. Hans CA, Braun P, Raisch J, Grune L, Reincke-Collon C (2019) Hierarchical distributed model predictive control of interconnected microgrids. IEEE Trans Sustain Energy 10(1):407–416
23. Sampathirao AK, Hans CA, Raisch J (2018) Two-stage consensus-based distributed MPC for interconnected microgrids, pp 1–8. arXiv:1810.03361
24. Baker K, Guo J, Hug G, Li X (2016) Distributed MPC for efficient coordination of storage and renewable energy sources across control areas. IEEE Trans Smart Grid 7(2):992–1001
25. Halvgaard R, Vandenberghe L, Poulsen NK, Madsen H, Jørgensen JB (2016) Distributed model predictive control for smart energy systems. IEEE Trans Smart Grid 7(3):1675–1682
26. Braun P, Grüne L, Kellett CM, Weller SR, Worthmann K (2016) A distributed optimization algorithm for the predictive control of smart grids. IEEE Trans Autom Control 61(12):3898–3911
27. Ananduta W, Nedić A, Ocampo-Martinez C (2021) Distributed augmented Lagrangian method for link-based resource sharing problems of multi-agent systems. IEEE Trans Autom Control (in press). Available online at https://doi.org/10.1109/TAC.2021.3092561
28. Ananduta W, Barreiro-Gomez J, Ocampo-Martinez C, Quijano N (2018) Resilient information-exchange protocol for distributed model predictive control schemes. In: Proceedings of the American Control Conference (ACC), pp 1286–1291
29. Ananduta W, Barreiro-Gomez J, Ocampo-Martinez C, Quijano N (2018) Mitigation of communication failures in distributed model predictive control strategies. IET Control Theory Appl 12(18):2507–2515
30. Ananduta W, Ocampo-Martinez C, Nedić A (2021) A distributed augmented Lagrangian method over stochastic network for economic dispatch of large-scale energy systems. IEEE Trans Sustain Energy. Available online at https://doi.org/10.1109/TSTE.2021.3073510
31. Ananduta W, Ocampo-Martinez C (2019) Decentralized energy management of power networks with distributed generation using periodical self-sufficient repartitioning approach. In: Proceedings of the American Control Conference (ACC), pp 3230–3235
32. Ananduta W, Ocampo-Martinez C (2021) Event-triggered partitioning for non-centralized predictive-control-based economic dispatch of interconnected microgrids. Automatica. Available online at. https://doi.org/10.1016/j.automatica.2021.109829
33. Ananduta W, Maestre J-M, Ocampo-Martinez C, Ishii H (2018) Resilient distributed energy management for systems of interconnected microgrids. In: Proceedings of the IEEE 57th Conference on Decision and Control (CDC), Miami, USA, pp 6159–6164
34. Ananduta W, Maestre JM, Ocampo-Martinez C, Ishii H (2019) A resilient approach for distributed MPC-based economic dispatch in interconnected microgrids. In: Proceedings of the European Control Conference, Naples, Italy, pp 891–896

35. Ananduta W, Maestre JM, Ocampo-Martinez C, Ishii H (2020) Resilient distributed model predictive control for energy management of interconnected microgrids. Optimal Control Appl Methods 41(1):146–169
36. Ananduta W, Pippia T, Ocampo-Martinez C, Sijs J, De Schutter B (2019) Online partitioning method for decentralized control of linear switching large-scale systems. J Franklin Inst 356(6):3290–3313

Chapter 2
Non-centralized MPC-Based Economic Dispatch

This chapter presents the energy management problem of large-scale energy networks and a non-centralized framework to solve it. First, Sect. 2.1 provides the model of the systems. Then, in Sect. 2.2, the mathematical formulation of the problem is stated. Section 2.3 presents a non-centralized scheme based on model predictive control (MPC) as the general framework considered in this thesis, whereas Sect. 2.4 introduces the benchmark case that is used when performing some numerical simulations. Finally, Sect. 2.5 summarizes this chapter by giving some remarks.

2.1 Model of Large-Scale Energy Systems

Let a large-scale energy system be represented by an undirected graph $\mathcal{G} = (\mathcal{N}, \mathcal{E})$, where $\mathcal{N} = \{1, 2, \ldots, n\}$ denotes the set of nodes and $\mathcal{E} \subseteq \mathcal{N} \times \mathcal{N}$ denotes the set of links that physically connect the nodes. In this system, each node might consume, store, or produce energy. Each of the capabilities is represented by a component, namely load, storage unit, and generation unit, respectively. Moreover, each pair of nodes that are connected, e.g., nodes i and j such that $\{i, j\} \in \mathcal{E}$, might also exchange energy between each other. Additionally, there might exist a third party, from which a node can import energy if the node is connected to it.

Beside classifying based on their capabilities, the components of the networks can be further divided into two groups: dispatchable and non-dispatchable units. This classification is relevant since, in an energy management problem, the objective is to compute the set points of the dispatchable units. Both generation units and loads can be either dispatchable or non-dispatchable, whereas storage units are dispatchable. However, for simplification of notations, it is considered that all energy generation units are dispatchable, whereas all loads are non-dispatchable and might be uncertain. Note that, the generality of the model is not lost since dispatchable loads can be considered as negative production of energy and, similarly, non-dispatchable

W. W. Ananduta, *Non-centralized Optimization-Based Control Schemes for Large-Scale Energy Systems*, Springer Theses,
https://doi.org/10.1007/978-3-030-89803-8_2

generation units can be considered as the components that consume negative energy. Finally, denote by $\mathcal{N}^{\mathrm{dg}} \subseteq \mathcal{N}$, $\mathcal{N}^{\mathrm{st}} \subseteq \mathcal{N}$, and $\mathcal{N}^{\mathrm{tp}} \subseteq \mathcal{N}$ the set of nodes that have dispatchable generation units, storage units, and are connected with a third party, respectively.

In a non-centralized control scheme, it is assumed that there exists m computational units (controllers), where $m \leq n$. These controllers cooperatively manage the operation of the system $\mathcal{G}$. In this regard, the system must be partitioned, where the definition of partitions is given as follows:

Definition 2.1 *(Non-overlapping partitions)* Let $\mathcal{M}_p$, for $p = 1, 2, \ldots, m$, be non-empty subsets of $\mathcal{N}$. Then, the set $\boldsymbol{\mathcal{M}} = \{\mathcal{M}_p : p = 1, 2, \ldots, m\}$ defines m non-overlapping partitions of graph $\mathcal{G} = (\mathcal{N}, \mathcal{E})$ if $\bigcup_{p=1}^{m} \mathcal{M}_p = \mathcal{N}$ and $\mathcal{M}_p \cap \mathcal{M}_q = \emptyset$, for any $\mathcal{M}_p, \mathcal{M}_q \in \boldsymbol{\mathcal{M}}$ and $p \neq q$. □

Note that the partitioning of the network can be done based on the network design, market design, or the energy contracts [1–3]. Moreover, a large-scale network might also be partitioned optimally. For instance, in [4–6], the authors propose optimization-based partitioning approaches for electrical networks. In this thesis, it is assumed that the system has been initially partitioned. Later in Chap. 7, the partitions are dynamically changed as a result of the proposed non-centralized scheme.

Now, suppose that at each discrete-time step $t \in \mathbb{Z}_{\geq 0}$, the system $\mathcal{G}$ is partitioned into m non-overlapping subsystems, defined by the set $\boldsymbol{\mathcal{M}}_t = \{\mathcal{M}_{p,t} : p = 1, 2, \ldots, m\}$ according to Definition 2.1. Each subsystem consists of a number of connected nodes and is considered as a supra-node. Since each subsystem contains some nodes, then each subsystem has similar features of the nodes. In particular, each subsystem might consume, store, or produce energy. Moreover, two neighboring subsystems might also exchange energy. However, different from the nodes, each subsystem has a computation unit/local controller that manages the operation of the nodes that belong to that subsystem. In this thesis, the notions of subsystem and agent are used interchangeably as an independent controllable entity.

The network of subsystems/supra-nodes is described by the undirected graph $\mathcal{G}_t^{\mathrm{s}} = (\mathcal{P}, \mathcal{E}_t^{\mathrm{s}})$. The set $\mathcal{P} = \{1, 2, \ldots, m\}$ denotes the index set of the subsystems and $\mathcal{E}_t^{\mathrm{s}} \subseteq \mathcal{P} \times \mathcal{P}$ denotes the set of links among the subsystems. Furthermore, if there exists at least one node in subsystem p connected to a node in subsystem q, then $\{p, q\} \in \mathcal{E}_t^{\mathrm{s}}$. Therefore, let $\mathcal{N}_{p,t}^{\mathrm{s}}$ denote the set of neighbors of subsystem $p \in \mathcal{P}$ at time step t, i.e., $\mathcal{N}_{p,t}^{\mathrm{s}} := \{q \in \mathcal{P} : \exists\{i, j\} \in \mathcal{E}, i \in \mathcal{M}_{p,t}, j \in \mathcal{M}_{q,t}, p \neq q\}$. Figure 2.1 illustrates how the network of supra-nodes of a small-scale energy system is formed. Additionally, notice that each node is associated with a subsystem and the association might change over time t. In this regard, let $\phi : \mathcal{N} \times \mathbb{Z} \rightarrow \mathcal{P}$ be the function that gives the subsystem index with which a node is associated at time step t, i.e., $\phi(i, t) := p$, where $p \in \{p \in \mathcal{P} : i \in \mathcal{M}_{p,t}\}$. Note that, for non-overlapping partitions at $t \in \mathbb{Z}_{\geq 0}$, the set $\{p \in \mathcal{P} : i \in \mathcal{M}_{p,t}\}$ has a unique element, for any $i \in \mathcal{N}$.

It is important to note that the definition of subsystems used in this thesis coincides with that of microgrids [2, 7]. With the presence of distributed generation and storage units, the microgrid concept is considered as the building blocks of a smart grid

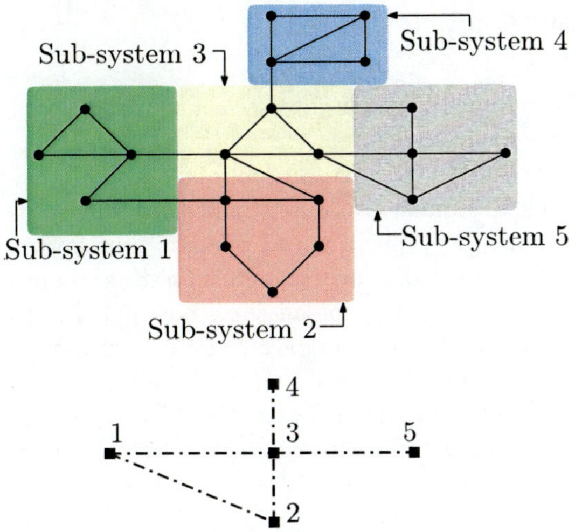

Fig. 2.1 An example of a small-scale energy network. Top figure shows graph $\mathcal{G}$ where the nodes in $\mathcal{N}$ are depicted as dots and the links in $\mathcal{E}$ are depicted as solid lines. The graph $\mathcal{G}$ is partitioned into 5 subsystems. Bottom figure shows graph $\mathcal{G}_t^{\mathrm{s}}$ where the subsystems (supra-nodes) in $\mathcal{P}$ are depicted as filled squares and the links in $\mathcal{E}_t^{\mathrm{s}}$ are depicted as dash-dotted lines

and has become a key to the development of non-centralized control approaches in energy systems [2]. Some literature, e.g., [8–10], starts the discussion of distributed or decentralized control approaches by assuming the existence of a network of microgrids. It is also worth mentioning that conceptually microgrids can operate in two modes: the grid-connected mode and the island mode, where it is not connected to the main grid. This microgrid feature will be revisited later in Chap. 8.

A network of energy prosumers (producer-consumer) and electrical distribution networks can be represented by the previously explained model. In a network of prosumers, e.g., networks of buildings, industrial sites, and residential areas that own small-scale generation and storage units, each node represents a prosumer. Therefore, in this case, each subsystem only consists of one node and the network of nodes and supra-nodes, i.e., $\mathcal{G}$ and $\mathcal{G}_t^{\mathrm{s}}$, are equivalent. Moreover, it also implies that the partitioning of the network raises naturally. Additionally, different forms of energy, e.g., not only electrical energy but also thermal energy, might exist and must be managed. Some examples of this type of networks are discussed in [3, 11, 12]. On the other hand, an electrical distribution network comprises a set of load busses might also have small-scale generation units, such as solar panels, wind turbines, or fuel-based generators, and storage units. According to the IEEE std 1547.4, such distribution networks can be partitioned into a number of microgrids to facilitate powerful control and operation infrastructure in the future distribution systems [6]. In this regard, the idea of having a network of microgrids is to improve robustness and self-healing features of the network [4, 6].

2.2 Economic Dispatch Problem

The objective of economic dispatch is to set economically optimal references for the controllable components such that the energy produced is equal to the energy consumed while at the same time the constraints of the system are satisfied. Therefore, an economic dispatch is formulated as a constrained optimization problem. Typically, the computed set points are defined over a specific time horizon. In a day-ahead planning problem, e.g., [13–15], hourly set points for the next one day are calculated. In some other economic dispatch problems, e.g., those that are discussed in [16–18], shorter sampling time and horizon might also be considered. Note that, since the control problem discussed in this thesis is a high-level one, i.e., computing the set points of the controllable components, it is assumed that these components have low-level controllers that control them such that the set points are met.

2.2.1 *Optimization Problem Formulation*

In order to state the economic dispatch problem considered in this thesis, first the decision variables are defined as follows. Recall that each node $i \in \mathcal{N}$ might obtain energy from a dispatchable generation unit, storage unit, or a third party (e.g. the main grid) depending on whether the node has those units and connected to the third party or not. Note that the set points are typically power, i.e., the amount of energy per time unit. Therefore, denote the *local decision* of node i at time step t by $u_{i,t} = \mathrm{col}(u_{i,t}^{\mathrm{st}}, u_{i,t}^{\mathrm{dg}}, u_{i,t}^{\mathrm{tp}}) \in \mathbb{R}^{n_i^{\mathrm{l}}}$, where $u_{i,t}^{\mathrm{st}} \in \mathbb{R}$, $u_{i,t}^{\mathrm{dg}} \in \mathbb{R}^{n_i^{\mathrm{dg}}}$, and $u_{i,t}^{\mathrm{tp}} \in \mathbb{R}_{\geq 0}$ are the set points of the power delivered from/to the storage unit, the power produced by the dispatchable generation units, and the power imported from the third party, respectively. Note that $n_i^{\mathrm{l}} = 2 + n_i^{\mathrm{dg}}$. Moreover, each node might also exchange power with its neighbor nodes. In this case, let $\mathcal{N}_i := \{j : \{i, j\} \in \mathcal{E}\}$ be the set of neighbors of node i. Then, the power exchanged between nodes i and $j \in \mathcal{N}_i$ is denoted by $v_{i,t}^j \in \mathbb{R}$. Note that $v_{i,t}^j$ is a decision of node i, whereas the decision of node j is denoted by $v_{j,t}^i$ and it must hold that

$$v_{i,t}^j + v_{j,t}^i = 0, \quad \forall j \in \mathcal{N}_i, \ \forall i \in \mathcal{N}. \tag{2.1}$$

For each $i \in \mathcal{N}$, collect the *coupled decisions* $v_{i,t}^j$, for all $j \in \mathcal{N}_i$, as a column vector $v_{i,t} := \mathrm{col}(\{v_{i,t}^j\}_{j \in \mathcal{N}_i})$. Finally, $d_{i,t} \in \mathbb{R}$ denotes the difference between the uncontrollable loads and the power generated by the non-dispatchable generation unit. If $d_{i,t} > 0$, then the aggregated uncontrollable load is larger than the power generated by the non-dispatchable unit. Therefore, for each node $i \in \mathcal{N}$, it also must hold that

$$\mathbb{1}_{n_i^{\mathrm{l}}}^\top u_{i,t} + \mathbb{1}_{|\mathcal{N}_i|}^\top v_{i,t} - d_{i,t} = 0, \tag{2.2}$$

which represents the local power balance equation of node i.

Now, denote the time horizon, over which the economic dispatch must be solved, by h and define an ordered set of time indices $\mathcal{T} := \{t, t+1, \ldots, t+h-1\}$. Furthermore, define the concatenated decision variables over the time horizon h with bold symbols, i.e., $\boldsymbol{u}_{i,t} := \mathrm{col}(\{u_{i,\tau}\}_{\tau \in \mathcal{T}}) \in \mathbb{R}^{n_i^l h}$, $\boldsymbol{v}_{i,t}^j := \mathrm{col}(\{v_{i,\tau}^j\}_{\tau \in \mathcal{T}}) \in \mathbb{R}^h$, and $\boldsymbol{d}_{i,t} = \mathrm{col}(\{d_{i,\tau}\}_{\tau \in \mathcal{T}}) \in \mathbb{R}^h$. Additionally, define that $\boldsymbol{v}_{i,t} = \mathrm{col}(\{\boldsymbol{v}_{i,t}^j\}_{j \in \mathcal{N}_i}) \in \mathbb{R}^{|\mathcal{N}_i|h}$. Then, the economic dispatch problem is stated as follows:

$$\underset{\{(\boldsymbol{u}_{i,t}, \boldsymbol{v}_{i,t})\}_{i \in \mathcal{N}}}{\text{minimize}} \sum_{i \in \mathcal{N}} \left(f_i^{\mathrm{l}}(\boldsymbol{u}_{i,t}) + f_i^{\mathrm{c}}(\boldsymbol{v}_{i,t}) \right) \tag{2.3a}$$

$$\text{s.t. } (\boldsymbol{u}_{i,t}, \boldsymbol{v}_{i,t}) \in \mathcal{L}_{i,t}, \ \forall i \in \mathcal{N}, \tag{2.3b}$$

$$\boldsymbol{v}_{i,t}^j + \boldsymbol{v}_{j,t}^i = 0, \quad \forall j \in \mathcal{N}_i, \ \forall i \in \mathcal{N}. \tag{2.3c}$$

In Problem (2.3), each node $i \in \mathcal{N}$ is assigned to a local cost function, which consists of two parts, f_i^{l} and f_i^{c}, which depend on $\boldsymbol{u}_{i,t}$ and $\boldsymbol{v}_{i,t}$, respectively. Moreover, in (2.3b), the decisions of each node $i \in \mathcal{N}$ are constrained by the local set $\mathcal{L}_{i,t}$, which is defined as follows: $\mathcal{L}_{i,t} = (\mathcal{U}_i \times \mathcal{V}_i) \cap \mathcal{W}_{i,t}$, where $\mathcal{U}_i \subset \mathbb{R}^{n_i^l h}$ is a constraint set of $\boldsymbol{u}_{i,t}$, $\mathcal{V}_i \subset \mathbb{R}^{|\mathcal{N}_i|h}$ is a constraint set of $\boldsymbol{v}_{i,t}$, and $\mathcal{W}_{i,t}$ is the set such that the power balance equation (2.2), for the whole time horizon, is satisfied, i.e.,

$$\mathcal{W}_{i,t} := \left\{ (\boldsymbol{u}, \boldsymbol{v}) \in \mathbb{R}^{(n_i^l + \mathcal{N}_i)h} : \left(I_h \otimes \mathbb{1}_{n_i^l}^\top \right) \boldsymbol{u} + \left(\mathbb{1}_{|\mathcal{N}_i|}^\top \otimes I_h \right) \boldsymbol{v} + \boldsymbol{d}_{i,t} = 0 \right\}. \tag{2.4}$$

Furthermore, the coupling constraints in (2.3c), which is obtained from (2.1), highlight the particularity of the problem structure. Due to these constraints, the problem can be considered as a link-based resource sharing problem. Finally, it is supposed that the following assumptions on Problem (2.3) hold.

Assumption 2.1 For each $i \in \mathcal{N}$, the functions $f_i^{\mathrm{l}} : \mathbb{R}^{n_i^l h} \to \mathbb{R}$ and $f_i^{\mathrm{c}} : \mathbb{R}^{|\mathcal{N}_i|h} \to \mathbb{R}$ are differentiable and convex. Moreover, $f_i^{\mathrm{l}}(\boldsymbol{u}_{i,t})$ is strongly convex with strong convexity constant σ_i. □

Assumption 2.2 For each $i \in \mathcal{N}$ and $t \in \mathbb{Z}_{\geq 0}$, $\mathcal{U}_i$ and $\mathcal{V}_i$ are non-empty, polyhedral, and compact whereas $\mathcal{L}_{i,t}$ is non-empty. □

Remark 2.1 By Assumption 2.2 and the definition of $\mathcal{W}_{i,t}$, the non-empty set $\mathcal{L}_{i,t}$, for each $i \in \mathcal{N}$ and $t \in \mathbb{Z}_{\geq 0}$, is polyhedral and compact. □

Assumption 2.3 The feasible set of Problem (2.3), for any $t \in \mathbb{Z}_{\geq 0}$, is non-empty. □

In the literature, e.g., [1, 10, 18, 19], a strongly convex quadratic cost function is typically considered as the cost function of producing energy, which is related to the local decisions $\boldsymbol{u}_{i,t}$, e.g.,

$$f_i^{\mathrm{l}}(\boldsymbol{u}_{i,t}) = \langle \boldsymbol{u}_{i,t}, Q_i \boldsymbol{u}_{i,t} \rangle + \langle \boldsymbol{q}_i, \boldsymbol{u}_{i,t} \rangle,$$

where a diagonal positive definite matrix $Q_i \in \mathbb{R}^{hn_i^l \times hn_i^l}$ and $\boldsymbol{q}_i \in \mathbb{R}^{hn_i^l}$ are constant. On the other hand, the cost function of exchanging energy can be either quadratic, as in [10], which considers a cost function that is based on the loss of energy when transfering power, or linear, as in [3], which considers the cost of trading energy among agents. In this thesis, the assumption on the cost function (Assumption 2.1) is more general and includes the typical cost functions considered in the literature.

The assumption on the local sets $\mathcal{U}_i$ and $\mathcal{V}_i$ (Assumption 2.2) is made since commonly the local and coupled decisions are constrained by the operating limitations of the components, which usually form a polyhedral set, whereas Assumption 2.3 is considered to ensure the existence of solutions of Problem 2.3 at each time step. Moreover, as seen in (2.4), the information of $d_{i,\tau}$, for all $\tau \in \mathcal{T}$, is required. However, since $\boldsymbol{d}_{i,t}$ is uncertain, one can only provide a forecast of $\boldsymbol{d}_{i,t}$, which is denoted by $\boldsymbol{d}^{\mathrm{f}}_{i,t}$. It is assumed that the forecast can be obtained, e.g., based on historical data. Now, denote $\boldsymbol{w}_{i,t}$ as the difference between $\boldsymbol{d}_{i,t}$ and $\boldsymbol{d}^{\mathrm{f}}_{i,t}$, i.e.,

$$\boldsymbol{d}_{i,t} = \boldsymbol{d}^{\mathrm{f}}_{i,t} + \boldsymbol{w}^{\mathrm{d}}_{i,t}, \quad \forall i \in \mathcal{N}. \tag{2.5}$$

Then, it is also assumed that $\boldsymbol{w}^{\mathrm{d}}_{i,t} = [w^{\mathrm{d}}_{i,\tau}]_{\tau\in\mathcal{T}}$ is bounded by $\boldsymbol{w}^{\mathrm{d,max}}_{i,t}$, i.e., $|\boldsymbol{w}^{\mathrm{d}}_{i,t}| \leq \boldsymbol{w}^{\mathrm{d,max}}_{i,t}$, where $\boldsymbol{w}^{\mathrm{d,max}}_{i,t}$, for each $i \in \mathcal{N}$ and $t \in \mathbb{Z}_{\geq 0}$, is known. Thus, one can consider the worst-case scenario of $\boldsymbol{d}_{i,t}$ in $\mathcal{W}_{i,t}$ to obtain robust decisions with respect to this uncertainty as follows:

$$\boldsymbol{d}_{i,t} = \boldsymbol{d}^{\mathrm{f}}_{i,t} + \boldsymbol{w}^{\mathrm{d,max}}_{i,t}, \quad \forall i \in \mathcal{N},\ \forall t \in \mathbb{Z}_{\geq 0}. \tag{2.6}$$

Therefore, the definition of $\mathcal{W}_{i,t}$ includes (2.6), implying the set $\mathcal{W}_{i,t}$ depends on the $\boldsymbol{w}^{\mathrm{d,max}}_{i,t}$. Moreover, note that in Chap. 7, a stochastic method is introduced to relax the assumption on $\boldsymbol{w}^{\mathrm{d}}_{i,t}$.

Now, typical constraints associated with the controllable components of the system and defining the local sets $\mathcal{U}_i$ and $\mathcal{V}_i$ are mentioned. These constraints also provide a model of the components considered in the economic dispatch problem (2.3). The operational constraints of the controllable components are, as follows:

1. *Dispatchable generation units.* The production capacity constraints of these units are stated as follows:

$$u_i^{\mathrm{dg,min}} \leq u_{i,\tau}^{\mathrm{dg}} \leq u_i^{\mathrm{dg,max}}, \quad \forall i \in \mathcal{N},\ \forall \tau \in \mathcal{T}, \tag{2.7}$$

   where the vector $u_i^{\mathrm{dg,min}}$ and $u_i^{\mathrm{dg,max}}$ denote the minimum and maximum power that can be generated by the dispatchable generation units in node i, respectively. Note that, for $i \notin \mathcal{N}^{\mathrm{dg}}$, $u_i^{\mathrm{dg,min}} = u_i^{\mathrm{dg,max}} = 0$.
2. *Storage units.* The capability of storing energy is modeled by discrete-time integrator dynamics, i.e.,

$$x_{i,\tau+1} = a_i x_{i,\tau} + b_i u^{\mathrm{st}}_{i,\tau}, \quad \forall i \in \mathcal{N}^{\mathrm{st}},\ \forall \tau \in \mathcal{T}, \tag{2.8}$$

   where x_i denotes the state of charge (SoC) of the storage unit. The scalar $a_i \in (0, 1]$ denotes the efficiency of the storage whereas $b_i = -\frac{T_s}{e_i^{\mathrm{cap}}}$, where T_s denotes the sampling time of the system and e_i^{cap} denotes the energy capacity of the storage unit. Moreover, the SoC and the power delivered to/from the storage are also constrained as follows:

$$x_i^{\min} \leq x_{i,\tau+1} \leq x_i^{\max}, \quad \forall i \in \mathcal{N}^{\text{st}}, \ \forall \tau \in \mathcal{T}, \tag{2.9}$$

$$-u_i^{\text{ch}} \leq u_{i,\tau}^{\text{dg}} \leq u_i^{\text{dh}}, \quad \forall i \in \mathcal{N}^{\text{st}}, \ \forall \tau \in \mathcal{T}, \tag{2.10}$$

$$u_{i,\tau}^{\text{dg}} = 0, \quad \forall i \notin \mathcal{N}^{\text{st}}, \ \forall \tau \in \mathcal{T}, \tag{2.11}$$

where $x_i^{\min}, x_i^{\max} \in [0, 1]$ denote the upper and lower limit of SoC whereas $u_i^{\text{ch}}, u_i^{\text{dh}} \in \mathbb{R}_{\geq 0}$ denote the maximum charging and discharging power of the storage.

3. *Imported power.* The amount of power that can be imported from a third party is limited as follows:

$$0 \leq u_{i,\tau}^{\text{tp}} \leq u_i^{\text{tp,max}}, \quad \forall i \in \mathcal{N}^{\text{tp}}, \ \forall \tau \in \mathcal{T}, \tag{2.12}$$

$$u_{i,\tau}^{\text{tp}} = 0, \quad \forall i \notin \mathcal{N}^{\text{tp}}, \ \forall \tau \in \mathcal{T}, \tag{2.13}$$

where $u_i^{\text{tp,max}}$ denotes the maximum power that can be imported.

4. *Exchanged power.* The amount of power that can be exchanged between two neighboring nodes might also be limited as follows:

$$-v_i^{j,\max} \leq v_{i,\tau}^j \leq v_i^{j,\max}, \quad \forall j \in \mathcal{N}_i, \ \forall i \in \mathcal{N}, \ \forall \tau \in \mathcal{T}, \tag{2.14}$$

where $v_i^{j,\max}$ is the maximum power that can be delivered to/from node i through link $\{i, j\}$.

Remark 2.2 One might consider additional constraints, e.g., those that are intertemporal. For instance, the rate of power produced by the generation units or the rate of power delivered to/from the storage units. With these constraints, Assumption 2.2 still holds. □

2.2.2 *Problem Reformulation Based on the Network of Supra-Nodes*

As discussed in Sect. 2.1, the network $\mathcal{G}$ is partitioned into m subsystems, each of which is managed by a local controller. Therefore, it is necessary to restate the economic dispatch problem (2.3) from the perspective of the network of subsystems/supra-nodes.

Proposition 2.1 *Suppose that, at time step t, the system $\mathcal{G}$ is partitioned into m non-overlapping subsystems, defined by $\boldsymbol{\mathcal{M}}_t = \{\mathcal{M}_{p,t} : p \in \mathcal{P}\}$, i.e., $\mathcal{M}_{p,t}$, for all $p \in \mathcal{P}$, satisfy Definition 2.1. Then, Problem (2.3) can be rewritten as follows:*

$$\underset{\{u_{i,t},v_{i,t}\}_{i\in\mathcal{N}}}{\text{minimize}} \sum_{p\in\mathcal{P}}\sum_{i\in\mathcal{M}_{p,t}} \left(f_i^{\mathrm{l}}(u_{i,t}) + f_i^{\mathrm{c}}(v_{i,t})\right) \tag{2.15a}$$

$$\text{s.t. } (u_{i,t}, v_{i,t}) \in \mathcal{L}_i,\ \forall i \in \mathcal{M}_{p,t},\ \forall p \in \mathcal{P}, \tag{2.15b}$$

$$v_{i,t}^j + v_{j,t}^i = 0,\quad \forall j \in \mathcal{N}_i \cap \mathcal{M}_{p,t},\ \forall i \in \mathcal{M}_{p,t},\ \forall p \in \mathcal{P}, \tag{2.15c}$$

$$v_{i,t}^j + v_{j,t}^i = 0,\quad \forall j \in \mathcal{N}_i \backslash \mathcal{M}_{p,t},\ \forall i \in \mathcal{M}_{p,t},\ \forall p \in \mathcal{P}. \tag{2.15d}$$

Proof Since $\mathcal{M}_{p,t}$, for all $p \in \mathcal{P}$, are non-overlapping, by Definition 2.1, it holds that $\mathcal{M}_{p,t} \cap \mathcal{M}_{q,t} = \emptyset$, for any $p \neq q$, where $p, q \in \mathcal{P}$, and $\bigcup_{p\in\mathcal{P}} \mathcal{M}_{p,t} = \mathcal{N}$. Thus, the cost function in (2.15a) are equivalent to the cost function in (2.3a), the constraints in (2.15b) are equivalent to the constraints in (2.3b). Moreover, the equality constraints in (2.3c) are decomposed into (2.15c) and (2.15d).

By rewriting the problem as in (2.15), it can be observed that the cost function is decomposable. In this regard, each subsystem can be assigned its own local cost function, i.e., $\sum_{i\in\mathcal{M}_{p,t}} \left(f_i^{\mathrm{l}}(u_{i,t}) + f_i^{\mathrm{c}}(v_{i,t})\right)$. Similarly, the constraints in (2.15b) and (2.15c) are also decomposable and each subsystem only consider its local constraints, i.e., $(u_{i,t}, v_{i,t}) \in \mathcal{L}_i$, for all $i \in \mathcal{M}_{p,t}$, and $v_{i,t}^j + v_{j,t}^i = 0$, for all $j \in \mathcal{N}_i \cap \mathcal{M}_{p,t}$ and $i \in \mathcal{M}_{p,t}$. Therefore, $u_{i,t}$, for all $i \in \mathcal{M}_{p,t}$, as well as $v_{i,t}^j$ and $v_{j,t}^i$, for all $j \in \mathcal{N}_i \cap \mathcal{M}_{p,t}$ and $i \in \mathcal{M}_{p,t}$, are local decisions of supra-node p. Finally, it is also observed that each equality constraint in (2.15d) couples two neighboring subsystems, i.e., subsystems $\phi(i,t)$ and $\phi(j,t)$, for each $j \in \mathcal{N}_i \backslash \mathcal{M}_{p,t}$, $i \in \mathcal{M}_{p,t}$, and $p \in \mathcal{P}$. Note that $\phi(j,t) \in \mathcal{N}_{\phi(i,t),t}^{\mathrm{s}}$ and, vice versa, $\phi(i,t) \in \mathcal{N}_{\phi(j,t),t}^{\mathrm{s}}$. In this regard, the decisions $v_{i,t}^j$, for all $j \in \mathcal{N}_i \backslash \mathcal{M}_{p,t}$ and $i \in \mathcal{M}_{p,t}$, are coupled decisions of supra-node p.

2.3 Non-centralized MPC Scheme

In this thesis, the MPC framework [20, 21], which uses the receding horizon principle, is applied. Therefore, at each time step, the economic dispatch problem over a fixed time horizon based on the measurement at that time step is solved and only the decisions at the first time instant are applied to the system. In general, an MPC-based economic dispatch algorithm consists of the steps detailed in Algorithm 2.1.

Algorithm 2.1 MPC-based economic dispatch

Iteration: For each $t \in \mathbb{Z}_{\geq 0}$,

1. Compute $u_{i,t}$ and $v_{i,t}$, for all $i \in \mathcal{N}$, by solving the optimization problem (2.15).
2. Implement the decisions of the current time step t, i.e., $u_{i,t}$ and $v_{i,t}$, for all $i \in \mathcal{N}$.
3. Measure the states $x_{i,t+1}$, for all $i \in \mathcal{N}^{\mathrm{st}}$.

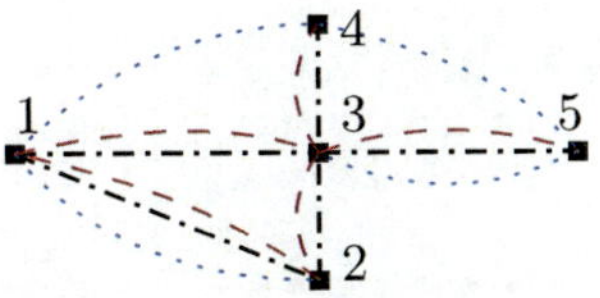

Fig. 2.2 The set of communication links at k_1, $\mathcal{E}^c(k_1)$, is depicted by dashed lines whereas the set of communication links at k_2, $\mathcal{E}^c(k_2)$, is depicted by dotted lines. The graph $\mathcal{G}_t^s$, which is fixed for all k, is shown by filled squared and dash-dotted lines

For step 1 of Algorithm 2.1, non-centralized methods are considered to solve the optimization problem (2.15). Note that a non-centralized approach to perform step 1 requires cooperation and communication among the controllers of the subsystems. In step 2, each subsystem must implement the decisions that have been computed in the previous step. In this regard, the compliance of all subsystems with respect to the computed decisions is required. Finally, in step 3, it is assumed that each subsystem can measure its states. The discussions in this thesis are focused on steps 1 and 2 of Algorithm 2.1. In particular, Chaps. 3–7 address non-centralized approaches to compute the decisions whereas Chap. 8 focuses on a non-compliance problem that might arise in step 2.

Finally, in regard to the non-centralized architecture, local controllers might need to exchange some information among each other over a communication network. The model of this network is given as follows. Let the undirected time-varying graph $\mathcal{G}^c(k) = (\mathcal{P}, \mathcal{E}^c(k))$ represents the communication network of local controllers where $\mathcal{E}^c(k)$ denotes the set of bidirectional communication links that interconnect the local controllers. Therefore, if $\{i, j\} \in \mathcal{E}^c(k)$, the local controllers of subsystems i and j can communicate and exchange information at discrete-time step k. Note that the time indices k and t are on a different time scale. While t defines the time step of the system, k defines the time step of the controllers or the algorithms, i.e., it represents the iteration step of the algorithms. In this regard, the sampling period that defines the discrete-time step t is assumed to be much larger than the sampling period that defines k. Additionally, to avoid confusion, the time indices t and k are written differently, i.e., t as a subscript and k in parentheses. An example of a time-varying communication network of the system shown in Fig. 2.1 is depicted in Fig. 2.2.

2.4 Benchmark Case

The numerical studies in some chapters are based on the PG&E 69-buss electrical distribution network, depicted in Fig. 2.3. Each bus is considered as a node, thus $n = 69$. The available load data of each bus is used as the maximum value of the non-dispatchable loads, which vary over time. Moreover, dispatchable and non-dispatchable generation units as well as storage units are added in some busses. The locations where these units are placed might vary from one study to another and will be specified later in the corresponding chapter. Nevertheless, an initial partition

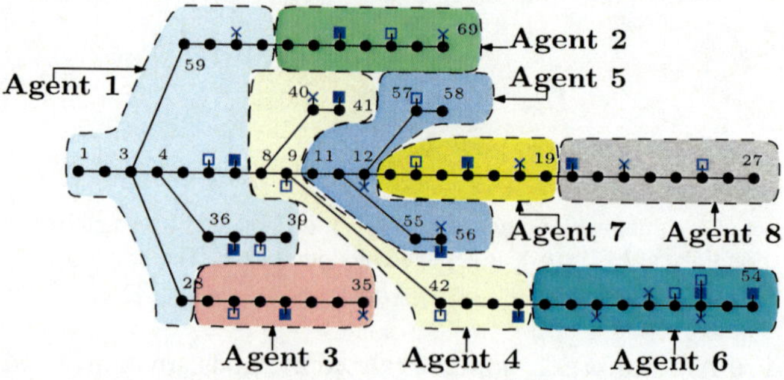

Fig. 2.3 The topology of the PG&E 69-bus distribution system and its 8-agent initial partition [4]. Squares indicate the distributed generation units, i.e., ■ and □ represent a renewable generation unit and a dispatchabale generator, respectively, whereas crosses, ×, indicate the storages

based on [4] and shown in Fig. 2.3 is used to form the supra-node network $\mathcal{G}^{\mathrm{s}}$. Based on this partitioning result, there are 8 interconnected subsystems.

In the economic dispatch problem of the benchmark case, a quadratic cost function of $f_i^{\mathrm{l}}(\boldsymbol{u}_i)$, for all $i \in \mathcal{N}$, in the form

$$f_i^{\mathrm{l}}(\boldsymbol{u}_{i,t}) = \boldsymbol{u}_{i,t}^{\top} Q_i^{\mathrm{l}} \boldsymbol{u}_{i,t} \tag{2.16}$$

is considered. Note that $Q_i^{\mathrm{l}} = I_h \otimes \mathrm{diag}(c_i^{\mathrm{st}}, c_i^{\mathrm{dg}}, c_i^{\mathrm{tp}})$, where c_i^{st}, c_i^{dg}, and c_i^{tp} denote the per-unit cost of power delivered to/from the storage unit, the per-unit cost of power generated by the dispatchable generator, and the per-unit cost of importing power from the third party. Moreover, by denoting c_i^{t} per-unit cost of power transferred by node i, the cost function $f_i^{\mathrm{c}}(\boldsymbol{v}_i)$, for all $i \in \mathcal{N}$, are linear in Chap. 6, as follows

$$f_i^{\mathrm{c}}(\boldsymbol{v}_{i,t}) = c_i^{\mathrm{t}} \mathbb{1}_{h|\mathcal{N}_i|}^{\top} \boldsymbol{v}_{i,t}, \tag{2.17}$$

whereas in Chaps. 7 and 8, they are quadratic, as follows:

$$f_i^{\mathrm{c}}(\boldsymbol{v}_{i,t}) = \boldsymbol{v}_{i,t}^{\top} Q_i^{\mathrm{c}} \boldsymbol{v}_{i,t}, \tag{2.18}$$

where $Q_i^{\mathrm{c}} = c_i^{\mathrm{t}} I_{h|\mathcal{N}_i|}$. In addition, the local constraint sets $\mathcal{U}_i$ and $\mathcal{V}_i$, for all $i \in \mathcal{N}$, are formed by (2.6)–(2.14).

2.5 Summary

This chapter presents the model of large-scale energy systems and their economic dispatch problem. General assumptions on the system and the problem are also specified. Moreover, the non-centralized MPC-based control framework is also introduced. Next, in Chap. 3, two distributed optimization methods that are designed to solve the economic dispatch problem are presented.

References

1. Kraning M, Chu E, Lavaei J, Boyd S (2014) Dynamic network energy management via proximal message passing. Found Trends Optim 1(2):73–126
2. Schwaegerl C, Tao L (2013) The microgrids concept. Wiley, pp 1–24
3. Sousa T, Soares T, Pinson P, Moret F, Baroche T, Sorin E (2019) Peer-to-peer and community-based markets: a comprehensive review. Renew Sustain Energy Rev 104:367–378
4. Arefifar SA, Mohamed YA-RI, El-Fouly THM (2012) Supply-adequacy-based optimal construction of microgrids in smart distribution systems. IEEE Trans Smart Grid 3(3):1491–1502
5. Arefifar SA, Mohamed YARI, EL-Fouly THM (2013) Optimum microgrid design for enhancing reliability and supply-security. IEEE Trans Smart Grid 4(3):1567–1575
6. Barani M, Aghaei J, Akbari MA, Niknam T, Farahmand, H, Korpås M (2018) Optimal partitioning of smart distribution systems into supply-sufficient microgrids. IEEE Trans Smart Grid
7. Olivares DE, Mehrizi-Sani A, Etemadi AH, Cañizares CA, Iravani R, Kazerani M, Hajimiragha AH, Gomis-Bellmunt O, Saeedifard M, Palma-Behnke R, Jimènez-Estèvez GA, Hatziargyriou ND (2014) Trends in microgrid control. IEEE Trans Smart Grid 5(4):1905–1919
8. Liang H, Choi BJ, Abdrabou A, Zhuang W, Shen XS (2012) Decentralized economic dispatch in microgrids via heterogeneous wireless networks. IEEE J Selected Areas Commun 30(6):1061–1074
9. Parisio A, Wiezorek C, Kyntàjà T, Elo J, Strunz K, Johansson KH (2017) Cooperative MPC-based energy management for networked microgrids. IEEE Trans Smart Grid 8(6):3066–3074
10. Hans CA, Braun P, Raisch J, Grune L, Reincke-Collon C (2019) Hierarchical distributed model predictive control of interconnected microgrids. IEEE Trans Sustain Energy 10(1):407–416
11. Sorin E, Bobo L, Pinson P (2019) Consensus-based approach to peer-to-peer electricity markets with product differentiation. IEEE Transactions on Power Systems 34(2):994–1004
12. Rostampour V, Keviczky T (2019) Probabilistic energy management for building climate comfort in smart thermal grids with seasonal storage systems. IEEE Transactions on Smart Grid 10(4):3687–3697
13. Garcia-Torres F, Bordons C (2015) Optimal economical schedule of hydrogen-based microgrids with hybrid storage using model predictive control. IEEE Trans Ind Electron 62(8):5195–5207
14. Atzeni I, Ordóñez LG, Scutari G, Palomar DP, Fonollosa JR (2013) Demand-side management via distributed energy generation and storage optimization. IEEE Trans Smart Grid 4(2):866–876
15. Capitanescu F, Martinez Ramos JL, Panciatici P, Kirschen D, Marano Marcolini A, Platbrood L, Wehenkel L (2011) State-of-the-art, challenges, and future trends in security constrained optimal power flow. Electric Power Syst Res 81(8):1731–1741
16. Garcia-Torres F, Valverde L, Bordons C (2016) Optimal load sharing of hydrogen-based microgrids with hybrid storage using model-predictive control. IEEE Trans Ind Electron 63(8):4919–4928
17. Parisio A, Rikos E, Glielmo L (2014) A model predictive control approach to microgrid operation optimization. IEEE Trans Control Syst Technol 22(5):1813–1827
18. Baker K, Guo J, Hug G, Li X (2016) Distributed MPC for efficient coordination of storage and renewable energy sources across control areas. IEEE Trans Smart Grid 7(2):992–1001
19. Molzahn DK, Dörfler F, Sandberg H, Low SH, Chakrabarti S, Baldick R, Lavaei J (2017) A survey of distributed optimization and control algorithms for electric power systems. IEEE Trans Smart Grid 8(6):2941–2962
20. Rawlings JB, Mayne DQ (2009) Model predictive control: theory and design. Nob Hill Publication, Madison, Wisconsin
21. Christofides PD, Scattolini R, Muñoz de la Peña D, Liu J (2013) Distributed model predictive control: a tutorial review and future research directions. Computers Chem Eng 51:21–41

Chapter 3
Distributed Augmented Lagrangian Methods

The discussion about non-centralized MPC schemes for economic dispatch is started by presenting two distributed optimization algorithms that solve Problem (2.15). The proposed methods are based on the augmented Lagrangian approach. First, in Sect. 3.1, a brief introduction about the augmented Lagrangian approach is presented. Then, in Sect. 3.2, a distributed algorithm based on this approach is designed and its convergence properties are stated. Section 3.3 discusses another distributed algorithm based on the ADMM, which also belongs to the class of the augmented Lagrangian methods. Similarly, the design and convergence analysis of the second algorithm is provided. Finally, some comparisons of the proposed algorithms and conclusions are drawn in Sect. 3.4.

3.1 Augmented Lagrangian Methods

Recall the network of subsystems $\mathcal{G}^{\mathrm{s}} = (\mathcal{P}, \mathcal{E}^{\mathrm{s}})$ and Problem (2.15) without time index t, as follows:

$$\underset{\{(\boldsymbol{u}_i,\boldsymbol{v}_i)\in\mathcal{L}_i\}_{i\in\mathcal{N}}}{\text{minimize}} \sum_{p\in\mathcal{P}} \sum_{i\in\mathcal{M}_p} \left(f_i^{\mathrm{l}}(\boldsymbol{u}_i) + f_i^{\mathrm{c}}(\boldsymbol{v}_i)\right) \tag{3.1a}$$

$$\text{s.t. } \boldsymbol{v}_i^j + \boldsymbol{v}_j^i = 0, \quad \forall j \in \mathcal{N}_i,\ \forall i \in \mathcal{M}_p,\ \forall p \in \mathcal{P}, \tag{3.1b}$$

where (2.15c) and (2.15d) are represented by (3.1b). Note that since the goal of this chapter is to show distributed methods that solve (2.15) at a fixed time step t, the time index t is dropped to simplify the notation. It is considered that the local decisions $\boldsymbol{u}_i$, for all $i \in \mathcal{M}_p$, are private and cannot be shared whereas the coupled decisions $\boldsymbol{v}_i$, for all $i \in \mathcal{M}_p$, can be shared. Problem (3.1) is a multi-agent optimization problem

W. W. Ananduta, *Non-centralized Optimization-Based Control Schemes for Large-Scale Energy Systems*, Springer Theses,
https://doi.org/10.1007/978-3-030-89803-8_3

where m computational units (agents) cooperatively optimize a separable convex cost function subject to convex local constraints and equality coupling constraints. The problem presented in (3.1) can be considered as a resource sharing problem, where the coupling exists in the links of the network. Moreover, it can also be reformulated as an extended monotropic optimization problem [1]. Nevertheless, the particular structure of the coupling constraints will be exploited to design scalable distributed methods.

The main challenge to solve (3.1) in a non-centralized manner is the existence of equality coupling constraints, which makes Problem (3.1) not trivially separable. In order to overcome this challenge and design a distributed algorithm, the Lagrange dual theory on convex optimization is considered [2]. The main idea of this concept is to relax the coupling constraints such that the relaxed problem is decomposable. In this regard, Lagrange multipliers associated with the coupling constraints are introduced. In the dual problem, these multipliers are maximized. For interested readers, this concept is extensively discussed in [3, 4]. Many distributed optimization methods, including those for energy management problems, e.g., [5–8], are developed based on solving the dual problem.

As previously mentioned, the problem considered is suitable to be decomposed using the Lagrange dual approach. Although it is possible to reformulate the problem into a consensus-based problem [9–11], the latter approach can become impractical when the number of agents is large since the information that must be exchanged is unnecessarily large. On the other hand, by employing a Lagrange dual-based method, the information necessary to be exchanged only depends on the neighboring agents. In order to deal with a larger class of cost functions, particularly those that are not necessarily strongly convex, the problem is augmented by introducing an auxiliary quadratic term which will be useful in the convergence analysis.

In the literature, there have been some distributed methods that are developed based on the augmented Lagrangian approach for different types of problems, including the extended monotropic ones. They include the ADAL method [12–14], diagonal quadratic approximation algorithm [15–17], and the ADMM [2, 11, 18–20]. The main difference among these algorithms is the technique that they use to decompose the augmented Lagrangian. In this chapter, two distributed optimization algorithms particularly designed to solve Problem (3.1) are presented. One algorithm is inspired by the ADAL method [12] whereas the other is an ADMM-based algorithm. Different from the distributed methods presented in the aforementioned literature, the proposed algorithms consider the partition of the decision variables and exploit the special structure of the coupling constraints to achieve scalability.

3.2 Distributed Augmented Lagrangian Algorithm

The distributed algorithm presented in this section is closely related to the ADAL method, discussed in [12, 14]. Similar to the ADAL method, some information from the neighbors is required in the local optimization step. Additionally, a convex combination step to update the primal variable is also necessary. Differently, in the proposed method, each agent only performs a convex combination step to update the coupled variables instead of all the decisions. Moreover, since a different augmented Lagrangian function than that considered in [12], the condition of the step size, which guarantees convergence, is also different.

3.2.1 Algorithm Design

Consider the augmented problem of (3.1) in the following form:

$$\underset{\{(\boldsymbol{u}_i,\boldsymbol{v}_i)\in\mathcal{L}_i\}_{i\in\mathcal{N}}}{\text{minimize}} \sum_{p\in\mathcal{P}}\sum_{i\in\mathcal{M}_p}\left(f_i^{\mathrm{l}}(\boldsymbol{u}_i)+f_i^{\mathrm{c}}(\boldsymbol{v}_i)+\sum_{j\in\mathcal{N}_i}\|\boldsymbol{v}_i^j+\boldsymbol{v}_j^i\|_2^2\right) \tag{3.2}$$
$$\text{s.t.}\ \ \boldsymbol{v}_i^j+\boldsymbol{v}_j^i=0,\quad \forall j\in\mathcal{N}_i,\ \forall i\in\mathcal{M}_p,\ \forall p\in\mathcal{P}.$$

The Lagrange dual approach is used to decompose Problem (3.2). To this end, denote the decisions of the whole network by $\boldsymbol{u}=\mathrm{col}(\{\boldsymbol{u}_i\}_{i\in\mathcal{N}})$ and $\boldsymbol{v}=\mathrm{col}(\{\boldsymbol{v}_i\}_{i\in\mathcal{N}})$ and introduce the Lagrangian of the augmented problem (3.2), denoted by $L_1(\boldsymbol{u},\boldsymbol{v},\boldsymbol{\lambda})$, as follows:

$$L_1(\boldsymbol{u},\boldsymbol{v},\boldsymbol{\lambda})=\sum_{i\in\mathcal{N}}\Big(f_i^{\mathrm{l}}(\boldsymbol{u}_i)+f_i^{\mathrm{c}}(\boldsymbol{v}_i)+\sum_{j\in\mathcal{N}_i}\Big(\langle\boldsymbol{\lambda}_i^j,\boldsymbol{v}_i^j+\boldsymbol{v}_j^i\rangle+\|\boldsymbol{v}_i^j+\boldsymbol{v}_j^i\|_2^2\Big)\Big), \tag{3.3}$$

where the coupled constraints in (3.1b) are relaxed and $\boldsymbol{\lambda}_i^j\in\mathbb{R}^h$, for all $j\in\mathcal{N}_i$ and $i\in\mathcal{N}$, are the Lagrange multipliers associated with them. Note that, for convenience, the Lagrange multipliers are compactly written as $\boldsymbol{\lambda}=\mathrm{col}(\{\boldsymbol{\lambda}_i\}_{i\in\mathcal{N}})$, where $\boldsymbol{\lambda}_i=\mathrm{col}(\{\boldsymbol{\lambda}_i^j\}_{j\in\mathcal{N}_i})$. Additionally, note also that $\mathcal{N}=\bigcup_{p\in\mathcal{P}}\mathcal{M}_p$ and $\mathcal{M}_p\neq\mathcal{M}_q$ if $p\neq q$ since the partitions are non-overlapping. Now, introduce the dual function, denoted by $g_1(\boldsymbol{\lambda})$ as follows:

$$g_1(\boldsymbol{\lambda})=\underset{\{(\boldsymbol{u}_i,\boldsymbol{v}_i)\in\mathcal{L}_i\}_{i\in\mathcal{N}}}{\text{minimize}}\ L_1(\boldsymbol{u},\boldsymbol{v},\boldsymbol{\lambda}). \tag{3.4}$$

The domain of $g_1(\boldsymbol{\lambda})$ is the entire space of $\boldsymbol{\lambda}$. It is also known from the duality theory that $g_1(\boldsymbol{\lambda})$ is concave and continuous. Therefore, based on the Weierstrass theorem, for any $\boldsymbol{\lambda}$, the value $g_1(\boldsymbol{\lambda})$ is finite.

The dual problem associated with Problem (3.2) is stated as follows:

$$\underset{\boldsymbol{\lambda}}{\text{maximize}}\ g_1(\boldsymbol{\lambda}). \tag{3.5}$$

Note that the dual optimal value is finite. Furthermore, the strong duality holds and the set of dual optimal points is non-empty since, in the primal problem (3.1), the cost function is convex and the constraints are linear [4, Proposition 5.2.1]. In other word, there exists a saddle point of the Lagrangian function $L_1(\boldsymbol{u}, \boldsymbol{v}, \boldsymbol{\lambda})$, i.e., a point $(\boldsymbol{u}^\star, \boldsymbol{v}^\star, \boldsymbol{\lambda}^\star) \in \prod_{i\in\mathcal{N}} \mathcal{L}_i \times \mathbb{R}^{\sum_{i\in\mathcal{N}} h|\mathcal{N}_i|}$ such that, for any $(\boldsymbol{u}, \boldsymbol{v}) \in \prod_{i\in\mathcal{N}} \mathcal{L}_i$ and $\boldsymbol{\lambda} \in \mathbb{R}^{\sum_{i\in\mathcal{N}} h|\mathcal{N}_i|}$, it holds that

$$L_1(\boldsymbol{u}^\star, \boldsymbol{v}^\star, \boldsymbol{\lambda}) \le L_1(\boldsymbol{u}^\star, \boldsymbol{v}^\star, \boldsymbol{\lambda}^\star) \le L_1(\boldsymbol{u}, \boldsymbol{v}, \boldsymbol{\lambda}^\star). \tag{3.6}$$

The saddle point relation implies that $(\boldsymbol{u}^\star, \boldsymbol{v}^\star)$ is a solution to Problem (3.1), while $\boldsymbol{\lambda}^\star$ is a dual optimal solution.

The dual function $g_1(\boldsymbol{\lambda})$ has separable constraints and all the terms in the Lagrangian function are also separable, except for the quadratic terms $\|\boldsymbol{v}_i^j + \boldsymbol{v}_j^i\|_2^2$, for all $j \in \mathcal{N}_i$ and $i \in \mathcal{N}$. In this regard, for each node $i \in \mathcal{N}$, the information from its neighbors will be used as a way to approximate the aforementioned quadratic terms and decompose $g_1(\boldsymbol{\lambda})$. For each node $i \in \mathcal{N}$, denote by $\tilde{\boldsymbol{v}}_j^i$ the information associated with $\boldsymbol{v}_j^i$ from neighbor $j \in \mathcal{N}_i$. Thus, the minimization on the right hand side of (3.4) is approximated by

$$\underset{\{(\boldsymbol{u}_i,\boldsymbol{v}_i)\in\mathcal{L}_i\}_{i\in\mathcal{N}}}{\text{minimize}} \sum_{p\in\mathcal{P}} \sum_{i\in\mathcal{M}_p} \left(f_i^{\mathrm{l}}(\boldsymbol{u}_i) + f_i^{\mathrm{c}}(\boldsymbol{v}_i) + \sum_{j\in\mathcal{N}_i} \left(\langle \boldsymbol{\lambda}_i^j, \boldsymbol{v}_i^j + \boldsymbol{v}_j^i \rangle + \|\boldsymbol{v}_i^j + \tilde{\boldsymbol{v}}_j^i\|_2^2 \right) \right). \tag{3.7}$$

It can be seen that the above minimization is decomposable. Therefore, assign a local minimization problem to each agent $p \in \mathcal{M}_p$, as follows:

$$\underset{\{(\boldsymbol{u}_i,\boldsymbol{v}_i)\in\mathcal{L}_i\}_{i\in\mathcal{M}_p}}{\text{minimize}} \sum_{i\in\mathcal{M}_p} \left(f_i^{\mathrm{l}}(\boldsymbol{u}_i) + f_i^{\mathrm{c}}(\boldsymbol{v}_i) + \sum_{j\in\mathcal{N}_i} \left(\langle \boldsymbol{\lambda}_i^j, \boldsymbol{v}_i^j + \boldsymbol{v}_j^i \rangle + \|\boldsymbol{v}_i^j + \tilde{\boldsymbol{v}}_j^i\|_2^2 \right) \right). \tag{3.8}$$

Notice that due to the strong convexity of $f_i^{\mathrm{l}}(\boldsymbol{u}_i)$ (Assumption 2.1) and the quadratic term $\|\boldsymbol{v}_i^j + \tilde{\boldsymbol{v}}_j^i\|_2^2$, there exists a unique solution to the minimization in (3.8).

Now, the distributed algorithm based on the augmented Lagrangian is ready to be presented. First, recall that $\phi(i, \cdot)$ is the index of the subsystem/agent with which node i is associated and let η_i^j, for all $j \in \mathcal{N}_i$ and $i \in \mathcal{N}$, denote the step sizes. Thus, the distributed method is shown in Algorithm 3.1. In step 1 of Algorithm 3.1, each agent updates the local decisions $\boldsymbol{u}_i(k+1)$, for all $i \in \mathcal{M}_p$, and auxiliary variables, which are denoted by $\hat{\boldsymbol{v}}_i(k)$, for all $i \in \mathcal{M}_p$, and used to update the shared decisions by solving (3.8), where $\tilde{\boldsymbol{v}}_j^i = \boldsymbol{v}_j^i(k)$, for all $j \in \mathcal{N}_i$ and $i \in \mathcal{M}_p$. Then, the update of $\boldsymbol{v}_i(k+1)$, for all $i \in \mathcal{M}_p$, by (3.10), where $\eta_i^j \in (0, 1)$, uses a convex combination of $\hat{\boldsymbol{v}}_i(k)$ and the value at the previous iteration $\boldsymbol{v}_i(k)$. Meanwhile, the dual variables are updated by (3.11), using the step size η_i^j, for all $j \in \mathcal{N}_i$. The choice of η_i^j will be discussed later when showing the convergence properties of Algorithm 3.1. Related to steps 3 and 5, $\boldsymbol{v}_j^i(k)$ and $\boldsymbol{\lambda}_j^i(k)$, for all $j \in \mathcal{N}_i \backslash \mathcal{M}_p$ and $i \in \mathcal{M}_p$, are decisions of agent $\phi(j, \cdot) \in \mathcal{N}_p^{\mathrm{s}}$. Therefore, agent p must receive information from agent $\phi(j, \cdot)$ and vice versa. On the other hand, $\boldsymbol{v}_j^i(k)$ and $\boldsymbol{\lambda}_j^i(k)$, for all $j \in \mathcal{N}_i \cap \mathcal{M}_p$ and $i \in \mathcal{M}_p$, are local decisions of agent p and are computed locally by agent p. The information exchange

Algorithm 3.1 Distributed augmented Lagrangian (DAL) method

Initialization: For each $i \in \mathcal{N}$, $\boldsymbol{v}_i(0) = \boldsymbol{v}_{i0} \in \mathbb{R}^{|\mathcal{N}_i|h}$ and $\boldsymbol{\lambda}_i(0) = \boldsymbol{\lambda}_{i0} \in \mathbb{R}^{|\mathcal{N}_i|h}$.
Iteration: For each agent $p \in \mathcal{P}$,

1. Update $\boldsymbol{u}_i(k+1)$ and $\hat{\boldsymbol{v}}_i(k)$, for all $i \in \mathcal{M}_p$, according to

$$\{(\boldsymbol{u}_i(k+1), \hat{\boldsymbol{v}}_i(k))\}_{i \in \mathcal{M}_p} = \arg \min_{\{(\boldsymbol{u}_i, \boldsymbol{v}_i) \in \mathcal{L}_i\}_{i \in \mathcal{M}_p}} \sum_{i \in \mathcal{M}_p} \Bigg(f_i^{\mathrm{l}}(\boldsymbol{u}_i) + f_i^{\mathrm{c}}(\boldsymbol{v}_i) + \sum_{j \in \mathcal{N}_i} \left(\langle \boldsymbol{\lambda}_i^j(k) + \boldsymbol{\lambda}_j^i(k), \boldsymbol{v}_i^j \rangle + \|\boldsymbol{v}_i^j + \boldsymbol{v}_j^i(k)\|_2^2 \right) \Bigg). \tag{3.9}$$

2. Update $\boldsymbol{v}_i^j(k+1)$, for all $j \in \mathcal{N}_i$ and $i \in \mathcal{M}_p$, as follows:

$$\boldsymbol{v}_i^j(k+1) = \eta_i^j \hat{\boldsymbol{v}}_i^j(k) + \left(1 - \eta_i^j\right) \boldsymbol{v}_i^j(k). \tag{3.10}$$

3. For each $j \in \mathcal{N}_i \backslash \mathcal{M}_p$ and $i \in \mathcal{M}_p$, send $\boldsymbol{v}_i^j(k+1)$ to and receive $\boldsymbol{v}_j^i(k+1)$ from agent $\phi(j, \cdot) \in \mathcal{N}_p^{\mathrm{s}}$.
4. Update the dual variables $\boldsymbol{\lambda}_i^j(k+1)$, for all $j \in \mathcal{N}_i$ and $i \in \mathcal{M}_p$, according to

$$\boldsymbol{\lambda}_i^j(k+1) = \boldsymbol{\lambda}_i^j(k) + \eta_i^j \left(\boldsymbol{v}_i^j(k+1) + \boldsymbol{v}_j^i(k+1) \right). \tag{3.11}$$

5. For each $j \in \mathcal{N}_i \backslash \mathcal{M}_p$ and $i \in \mathcal{M}_p$, send $\boldsymbol{\lambda}_i^j(k+1)$ to and receive $\boldsymbol{\lambda}_j^i(k+1)$ from agent $\phi(j, \cdot) \in \mathcal{N}_p^{\mathrm{s}}$.

---

requirement between neighboring subsystems implicitly provides an assumption on the communication network $\mathcal{G}^{\mathrm{c}}(k)$, as follows.

Assumption 3.1 The communication network $\mathcal{G}^{\mathrm{c}}(k)$ is static, i.e., $\mathcal{G}^{\mathrm{c}}(k) = (\mathcal{P}, \mathcal{E}^{\mathrm{c}}(k))$, where $\mathcal{E}^{\mathrm{c}}(k) = \mathcal{E}^{\mathrm{s}}$, for all $k \in \mathbb{Z}_{\geq 0}$. □

3.2.2 *Convergence Analysis*

In this section, the convergence of Algorithm 3.1 is shown. Prior to showing the main convergence results, some intermediate results in Lemmas 3.1–3.3 are established. Lemma 3.1 provides an estimate obtained from the optimality conditions of the local minimization problems (3.8) and the saddle point inequalities in (3.6). Lemma 3.2 manipulates this estimate and Lemma 3.3 uses it to provide a sufficient condition of the step sizes η_i^j, for all $j \in \mathcal{N}_i$ and $i \in \mathcal{N}$, such that the Lyapunov function, denoted by $V_1(k)$ and defined later, is monotonically non-increasing under Algorithm 3.1. Finally, the convergence of the sequence produced by Algorithm 3.1 is shown in Theorem 3.1 based on Lemma 3.3.

First, due to Assumption 2.1, recall some properties of convex and strongly convex differentiable functions, which will be used in the analysis.

Property 3.1 *(Convexity) A differentiable function $f : \mathbb{R}^n \to \mathbb{R}$ is convex, if, for any $x, y \in \mathbb{R}^n$, it holds that*

$$f(y) - f(x) \geq \langle \nabla f(x), y - x \rangle.$$

Property 3.2 *(Strong convexity [21, Theorem 5.24.iii]) A differentiable function $f : \mathbb{R}^n \to \mathbb{R}$ is strongly convex with strong convexity constant σ, if, for any $x, y \in \mathbb{R}^n$, it holds that*

$$\langle \nabla f(y) - \nabla f(x), y - x \rangle \geq \sigma \|y - x\|_2.$$

Now, the first building block to show the convergence of Algorithm 3.1 is stated.

Lemma 3.1 *Let Assumptions 2.1-2.3 hold. Furthermore, let $(\boldsymbol{u}_i(k+1), \hat{\boldsymbol{v}}_i(k))$, for each $i \in \mathcal{M}_p$, be the attainer of the local optimization in (3.9) and $(\boldsymbol{u}^\star, \boldsymbol{v}^\star, \boldsymbol{\lambda}^\star)$ be a saddle point of $L_1(\boldsymbol{u}, \boldsymbol{v}, \boldsymbol{\lambda})$ as defined in (3.3). Then, it holds that*

$$\begin{aligned} 0 \leq \sum_{i \in \mathcal{N}} \Bigg(&-\sigma_i \|\boldsymbol{u}_i(k+1) - \boldsymbol{u}_i^\star\|_2^2 + \sum_{j \in \mathcal{N}_i} \langle \boldsymbol{\lambda}_i^{j\star} - \boldsymbol{\lambda}_i^j(k), \hat{\boldsymbol{v}}_i^j(k) + \hat{\boldsymbol{v}}_j^i(k) \rangle \\ &- \sum_{j \in \mathcal{N}_i} \Big(\|\hat{\boldsymbol{v}}_i^j(k) + \hat{\boldsymbol{v}}_j^i(k)\|_2^2 + 2\langle \boldsymbol{v}_j^i(k) - \hat{\boldsymbol{v}}_j^i(k), \hat{\boldsymbol{v}}_i^j(k) - \boldsymbol{v}_i^{j\star} \rangle \Big) \Bigg). \end{aligned} \tag{3.12}$$

Proof Since $(\boldsymbol{u}_i^\star, \boldsymbol{v}_i^\star) \in \mathcal{L}_i$, for each $i \in \mathcal{M}_p$, the optimality condition [22, Theorem 20] of the local optimization in (3.9) yields the following relation:

$$\begin{aligned} 0 \leq \sum_{i \in \mathcal{M}_p} \Bigg(&\langle \nabla f_i^{\mathrm{l}}(\boldsymbol{u}_i(k+1)), \boldsymbol{u}_i^\star - \boldsymbol{u}_i(k+1) \rangle + \langle \nabla f_i^{\mathrm{c}}(\hat{\boldsymbol{v}}_i(k)), \boldsymbol{v}_i^\star - \hat{\boldsymbol{v}}_i(k) \rangle \\ &+ \sum_{j \in \mathcal{N}_i} \Big(\langle \boldsymbol{\lambda}_i^j(k) + \boldsymbol{\lambda}_j^i(k), \boldsymbol{v}_i^{j\star} - \hat{\boldsymbol{v}}_i^j(k) \rangle + 2\langle \hat{\boldsymbol{v}}_i^j(k) + \boldsymbol{v}_j^i(k), \boldsymbol{v}_i^{j\star} - \hat{\boldsymbol{v}}_i^j(k) \rangle \Big) \Bigg). \end{aligned} \tag{3.13}$$

Moreover, $(\boldsymbol{u}_i(k+1), \hat{\boldsymbol{v}}_i(k))$, for each $i \in \mathcal{M}_p$, is also locally feasible, i.e., $(\boldsymbol{u}_i(k+1), \hat{\boldsymbol{v}}_i(k)) \in \mathcal{L}_i$. Now, consider the second inequality in (3.6), which implies that

$$(\boldsymbol{u}^\star, \boldsymbol{v}^\star) = \arg \min_{\{(\boldsymbol{u}_i, \boldsymbol{v}_i) \in \mathcal{L}_i\}_{i \in \mathcal{N}}} L(\boldsymbol{u}, \boldsymbol{v}, \boldsymbol{\lambda}^\star).$$

Based on the optimality condition of the above minimization and the fact that $(\boldsymbol{u}_i(k+1), \hat{\boldsymbol{v}}_i(k)) \in \mathcal{L}_i$, it follows that

$$\begin{aligned} 0 \leq \sum_{i \in \mathcal{N}} \Bigg(&\langle \nabla f_i^{\mathrm{l}}(\boldsymbol{u}_i^\star), \boldsymbol{u}_i(k+1) - \boldsymbol{u}_i^\star \rangle + \langle \nabla f_i^{\mathrm{c}}(\boldsymbol{v}_i^\star), \hat{\boldsymbol{v}}_i(k) - \boldsymbol{v}_i^\star \rangle \\ &+ \sum_{j \in \mathcal{N}_i} \Big(\langle \boldsymbol{\lambda}_i^{j\star} + \boldsymbol{\lambda}_j^{i\star}, \hat{\boldsymbol{v}}_i^j(k) - \boldsymbol{v}_i^{j\star} \rangle + 4\langle \boldsymbol{v}_i^{j\star} + \boldsymbol{v}_j^{i\star}, \hat{\boldsymbol{v}}_i^j(k) - \boldsymbol{v}_i^{j\star} \rangle \Big) \Bigg), \end{aligned} \tag{3.14}$$

where the last term on the right-hand side of the inequality, i.e., $\sum_{i\in\mathcal{N}}\sum_{j\in\mathcal{N}_i} 4\langle \boldsymbol{v}_i^{j\star} + \boldsymbol{v}_j^{i\star}, \hat{\boldsymbol{v}}_i^j(k) - \boldsymbol{v}_i^{j\star}\rangle$, is equal to 0 since $\boldsymbol{v}_i^{j\star} + \boldsymbol{v}_j^{i\star} = 0$, for all $j \in \mathcal{N}_i$ and $i \in \mathcal{N}$. By summing up (3.13) over all agents in $\mathcal{P}$ and combining with (3.14), it holds that

$$
\begin{aligned}
0 \le \sum_{i\in\mathcal{N}} \Big(& \langle \nabla f_i^{\mathrm{l}}(\boldsymbol{u}_i^\star) - \nabla f_i^{\mathrm{l}}(\boldsymbol{u}_i(k+1)), \boldsymbol{u}_i(k+1) - \boldsymbol{u}_i^\star\rangle \\
& + \langle \nabla f_i^{\mathrm{c}}(\boldsymbol{v}_i^\star) - \nabla f_i^{\mathrm{c}}(\hat{\boldsymbol{v}}_i(k)), \hat{\boldsymbol{v}}_i(k) - \boldsymbol{v}_i^\star\rangle \\
& + \sum_{j\in\mathcal{N}_i} \langle \boldsymbol{\lambda}_i^{j\star} + \boldsymbol{\lambda}_j^{i\star} - \boldsymbol{\lambda}_i^j(k) - \boldsymbol{\lambda}_j^i(k), \hat{\boldsymbol{v}}_i^j(k) - \boldsymbol{v}_i^{j\star}\rangle \\
& -2\sum_{j\in\mathcal{N}_i} \langle \hat{\boldsymbol{v}}_i^j(k) + \boldsymbol{v}_j^i(k), \hat{\boldsymbol{v}}_i^j(k) - \boldsymbol{v}_i^{j\star}\rangle \Big).
\end{aligned}
\tag{3.15}
$$

For each $i \in \mathcal{N}$, the convexity of $f_i^{\mathrm{c}}(\boldsymbol{v}_i)$ (c.f. Property 3.1) gives $\langle \nabla f_i^{\mathrm{c}}(\boldsymbol{v}_i^\star), \hat{\boldsymbol{v}}_i(k) - \boldsymbol{v}_i^\star\rangle \le f_i^{\mathrm{c}}(\hat{\boldsymbol{v}}_i(k)) - f_i^{\mathrm{c}}(\boldsymbol{v}_i^\star)$ and $\langle \nabla f_i^{\mathrm{c}}(\hat{\boldsymbol{v}}_i(k)), \boldsymbol{v}_i^\star - \hat{\boldsymbol{v}}_i(k)\rangle \le f_i^{\mathrm{c}}(\boldsymbol{v}_i^\star) - f_i^{\mathrm{c}}(\hat{\boldsymbol{v}}_i(k))$ and the strong convexity of $f_i^{\mathrm{l}}(\boldsymbol{u}_i)$ (c.f. Property 3.2) gives $\langle \nabla f_i^{\mathrm{l}}(\boldsymbol{u}_i^\star) - \nabla f_i^{\mathrm{l}}(\boldsymbol{u}_i(k+1)), \boldsymbol{u}_i(k+1) - \boldsymbol{u}_i^\star\rangle \le -\sigma_i\|\boldsymbol{u}_i(k+1) - \boldsymbol{u}_i^\star\|_2^2$. Applying these relations to (3.15), it follows that

$$
\begin{aligned}
0 &\le \sum_{i\in\mathcal{N}} \Big(-\sigma_i\|\boldsymbol{u}_i(k+1) - \boldsymbol{u}_i^\star\|_2^2 + \sum_{j\in\mathcal{N}_i} \langle \boldsymbol{\lambda}_i^{j\star} + \boldsymbol{\lambda}_j^{i\star} - \boldsymbol{\lambda}_i^j(k) - \boldsymbol{\lambda}_j^i(k), \hat{\boldsymbol{v}}_i^j(k) - \boldsymbol{v}_i^{j\star}\rangle \\
&\quad -2\sum_{j\in\mathcal{N}_i} \langle \hat{\boldsymbol{v}}_i^j(k) + \boldsymbol{v}_j^i(k), \hat{\boldsymbol{v}}_i^j(k) - \boldsymbol{v}_i^{j\star}\rangle \Big) \\
&= \sum_{i\in\mathcal{N}} \Big(-\sigma_i\|\boldsymbol{u}_i(k+1) - \boldsymbol{u}_i^\star\|_2^2 + \sum_{j\in\mathcal{N}_i} \langle \boldsymbol{\lambda}_i^{j\star} + \boldsymbol{\lambda}_j^{i\star} - \boldsymbol{\lambda}_i^j(k) - \boldsymbol{\lambda}_j^i(k), \hat{\boldsymbol{v}}_i^j(k) - \boldsymbol{v}_i^{j\star}\rangle \\
&\quad -2\sum_{j\in\mathcal{N}_i} \langle \hat{\boldsymbol{v}}_i^j(k) + \hat{\boldsymbol{v}}_j^i(k), \hat{\boldsymbol{v}}_i^j(k) - \boldsymbol{v}_i^{j\star}\rangle + \langle \boldsymbol{v}_j^i(k) - \hat{\boldsymbol{v}}_j^i(k), \hat{\boldsymbol{v}}_i^j(k) - \boldsymbol{v}_i^{j\star}\rangle \Big),
\end{aligned}
\tag{3.16}
$$

where the equality is obtained by adding the term $\sum_{i\in\mathcal{N}}\sum_{j\in\mathcal{N}_i} 2\langle \hat{\boldsymbol{v}}_j^i(k) - \hat{\boldsymbol{v}}_j^i(k), \hat{\boldsymbol{v}}_i^j(k) - \boldsymbol{v}_i^{j\star}\rangle = 0$. Now, manipulate the second term on the right-hand side of the equality, i.e.,

$$
\sum_{i\in\mathcal{N}}\sum_{j\in\mathcal{N}_i} \langle \boldsymbol{\lambda}_i^{j\star} + \boldsymbol{\lambda}_j^{i\star} - \boldsymbol{\lambda}_i^j(k) - \boldsymbol{\lambda}_j^i(k), \hat{\boldsymbol{v}}_i^j(k) - \boldsymbol{v}_i^{j\star}\rangle.
$$

By considering the summation over all links and since at each link there exist two inner products associated with both agents coupled by that link, that term is equivalent to

$$
\begin{aligned}
&\sum_{\{i,j\}\in\mathcal{E}} \Big(\langle \boldsymbol{\lambda}_i^{j\star} + \boldsymbol{\lambda}_j^{i\star} - \boldsymbol{\lambda}_i^j(k) - \boldsymbol{\lambda}_j^i(k), \hat{\boldsymbol{v}}_i^j(k) - \boldsymbol{v}_i^{j\star} \rangle \\
&\qquad + \langle \boldsymbol{\lambda}_j^{i\star} + \boldsymbol{\lambda}_i^{j\star} - \boldsymbol{\lambda}_j^i(k) - \boldsymbol{\lambda}_i^j(k), \hat{\boldsymbol{v}}_j^i(k) - \boldsymbol{v}_j^{i\star} \rangle \Big) \\
&\quad = \sum_{\{i,j\}\in\mathcal{E}} \langle \boldsymbol{\lambda}_i^{j\star} + \boldsymbol{\lambda}_j^{i\star} - \boldsymbol{\lambda}_i^j(k) - \boldsymbol{\lambda}_j^i(k), \hat{\boldsymbol{v}}_i^j(k) + \hat{\boldsymbol{v}}_j^i(k) \rangle \\
&\quad = \sum_{i\in\mathcal{N}} \sum_{j\in\mathcal{N}_i} \langle \boldsymbol{\lambda}_i^{j\star} - \boldsymbol{\lambda}_i^j(k), \hat{\boldsymbol{v}}_i^j(k) + \hat{\boldsymbol{v}}_j^i(k) \rangle,
\end{aligned}
$$

where the first equality is obtained by using the fact that $\boldsymbol{v}_i^{j\star} + \boldsymbol{v}_j^{i\star} = 0$. Using similar analysis, the third term on the right-hand side of the equality in (3.16), i.e., $2\sum_{i\in\mathcal{N}}\sum_{j\in\mathcal{N}_i}\langle \hat{\boldsymbol{v}}_i^j(k) + \hat{\boldsymbol{v}}_j^i(k), \hat{\boldsymbol{v}}_i^j(k) - \boldsymbol{v}_i^{j\star}\rangle$ is equivalent to

$$
\sum_{\{i,j\}\in\mathcal{E}} 2\langle \hat{\boldsymbol{v}}_i^j(k) + \hat{\boldsymbol{v}}_j^i(k), \hat{\boldsymbol{v}}_i^j(k) + \hat{\boldsymbol{v}}_j^i(k) \rangle = \sum_{i\in\mathcal{N}} \sum_{j\in\mathcal{N}_i} \|\hat{\boldsymbol{v}}_i^j(k) + \hat{\boldsymbol{v}}_j^i(k)\|_2^2.
$$

Thus, the desired inequality (3.12) is obtained.

Then, a useful estimate is derived from the result obtained in Lemma 3.1. In this regard, define the auxiliary variables, $\tilde{\boldsymbol{\lambda}}_i^j(k)$, for all $j \in \mathcal{N}_i$ and $i \in \mathcal{N}$, as follows:

$$
\tilde{\boldsymbol{\lambda}}_i^j(k) = \boldsymbol{\lambda}_i^j(k) + (1 - \eta_i^j)(\boldsymbol{v}_i^j(k) + \boldsymbol{v}_j^i(k)). \tag{3.17}
$$

Lemma 3.2 *Let Assumptions 2.1–2.3 hold. Furthermore, let* $(\boldsymbol{u}_i(k+1), \hat{\boldsymbol{v}}_i(k))$*, for all* $i \in \mathcal{M}_p$*, be the attainer of the local optimization in (3.9) and* $(\boldsymbol{u}^\star, \boldsymbol{v}^\star, \boldsymbol{\lambda}^\star)$ *be a saddle point of* $L_1(\boldsymbol{u}, \boldsymbol{v}, \boldsymbol{\lambda})$ *as defined in (3.3). Furthermore, let* $\tilde{\boldsymbol{\lambda}}_i^j(k)$*, for all* $j \in \mathcal{N}_i$ *and* $i \in \mathcal{N}$*, be defined as in (3.17). Then, it holds that*

$$
\begin{aligned}
&\sum_{i\in\mathcal{N}} \left(\sum_{j\in\mathcal{N}_i} \langle \tilde{\boldsymbol{\lambda}}_i^j(k) - \boldsymbol{\lambda}_i^{j\star}, \hat{\boldsymbol{v}}_i^j(k) + \hat{\boldsymbol{v}}_j^i(k) \rangle + 2\langle \hat{\boldsymbol{v}}_i^j(k) - \boldsymbol{v}_i^j(k), \boldsymbol{v}_i^j(k) - \boldsymbol{v}_i^{j\star} \rangle \right) \\
&\le -\sum_{i\in\mathcal{N}} \Bigg(\sigma_i \|\boldsymbol{u}_i(k+1) - \boldsymbol{u}_i^\star\|_2^2 + \sum_{j\in\mathcal{N}_i} \frac{3}{2} \|\hat{\boldsymbol{v}}_i^j(k) - \boldsymbol{v}_i^j(k)\|_2^2 \\
&\qquad + \sum_{j\in\mathcal{N}_i} \frac{\eta_i^j + \eta_j^i - (\eta_i^j + \eta_j^i)^2}{2} \|\hat{\boldsymbol{v}}_i^j(k) + \hat{\boldsymbol{v}}_j^i(k)\|_2^2 \Bigg).
\end{aligned} \tag{3.18}
$$

Proof The inequality in (3.18) is obtained by algebraically manipulating the result in Lemma 3.1. The complete proof is given in Appendix A.1.

As the next building block to show the convergence result, define a Lyapunov function, denoted by $V_1(k)$. For any given saddle point of $L_1(\boldsymbol{u}, \boldsymbol{v}, \boldsymbol{\lambda})$ (see (3.3)), denoted by $(\boldsymbol{u}^\star, \boldsymbol{v}^\star, \boldsymbol{\lambda}^\star)$, $V_1(k)$ can be constructed as follows:

$$V_1(k) = \|\boldsymbol{v}(k) - \boldsymbol{v}^\star\|_H^2 + \frac{1}{2}\|\tilde{\boldsymbol{\lambda}}(k) - \boldsymbol{\lambda}^\star\|_H^2, \tag{3.19}$$

where $\tilde{\boldsymbol{\lambda}} = \mathrm{col}(\{\tilde{\boldsymbol{\lambda}}_i(k)\}_{i\in\mathcal{N}})$, $\tilde{\boldsymbol{\lambda}}_i(k) = \mathrm{col}(\{\tilde{\boldsymbol{\lambda}}_i^j(k)\}_{j\in\mathcal{N}_i})$, $\tilde{\boldsymbol{\lambda}}_i^j(k)$ is defined in (3.17), $H = \mathrm{blkdiag}(\{H_i\}_{i\in\mathcal{N}})$ and $H_i = \mathrm{blkdiag}(\{(\eta_i^j)^{-1}I_h\}_{j\in\mathcal{N}_i})$, for all $i \in \mathcal{N}$. In the next lemma, it is shown that $\{V_1(k)\}$ is monotonically non-increasing under Algorithm 3.1 with a certain condition of the step sizes η_i^j, for all $j \in \mathcal{N}_i$ and $i \in \mathcal{N}$. Moreover, an upper bound of $V_1(k+1) - V_1(k)$, for any $k \in \mathbb{Z}_{\geq 0}$, that will be used in the main theorems is also obtained.

Lemma 3.3 *Let Assumptions 2.1-2.3 and 3.1 hold. Let the sequence* $\{(\boldsymbol{u}(k), \boldsymbol{v}(k), \hat{\boldsymbol{v}}(k), \boldsymbol{\lambda}(k))\}$ *be generated by Algorithm 3.1 and* $(\boldsymbol{u}^\star, \boldsymbol{v}^\star, \boldsymbol{\lambda}^\star)$ *be a saddle point of* $L_1(\boldsymbol{u}, \boldsymbol{v}, \boldsymbol{\lambda})$ *as defined in (3.3). Furthermore, let* $V_1(k)$ *be defined in (3.19). If* $\eta_i^j = \eta_j^i = \eta_{ij} \in \left(0, \frac{1}{4}\right)$, *for all* $j \in \mathcal{N}_i$ *and* $i \in \mathcal{N}$, *then* $\{V_1(k)\}$ *is a monotonically non-increasing sequence and the following inequality holds:*

$$\begin{aligned} V_1(k+1) - V_1(k) \leq &- \sum_{i\in\mathcal{N}} \sigma_i \|\boldsymbol{u}_i(k+1) - \boldsymbol{u}_i^\star\|_2^2 \\ &- \sum_{i\in\mathcal{N}}\sum_{j\in\mathcal{N}_i} \left(\frac{3}{2} - \eta_{ij}\right) \|\hat{\boldsymbol{v}}_i^j(k) - \boldsymbol{v}_i^j(k)\|_2^2 \\ &- \sum_{i\in\mathcal{N}}\sum_{j\in\mathcal{N}_i} \frac{\eta_{ij} - (2\eta_{ij})^2}{2} \|\hat{\boldsymbol{v}}_i^j(k) + \hat{\boldsymbol{v}}_j^i(k)\|_2^2. \end{aligned} \tag{3.20}$$

□

Proof Firstly, notice that $\tilde{\boldsymbol{\lambda}}_i^j(k+1)$ can be expressed as follows:

$$\begin{aligned} \tilde{\boldsymbol{\lambda}}_i^j(k+1) &= \boldsymbol{\lambda}_i^j(k+1) + (1 - \eta_i^j)(\boldsymbol{v}_i^j(k+1) + \boldsymbol{v}_j^i(k+1)) \\ &= \boldsymbol{\lambda}_i^j(k) + \eta_i^j\Big(\boldsymbol{v}_i^j(k+1) + \boldsymbol{v}_j^i(k+1)\Big) + (1 - \eta_i^j)(\boldsymbol{v}_i^j(k+1) + \boldsymbol{v}_j^i(k+1)) \\ &= \boldsymbol{\lambda}_i^j(k) + \Big(\boldsymbol{v}_i^j(k+1) + \boldsymbol{v}_j^i(k+1)\Big) \\ &= \boldsymbol{\lambda}_i^j(k) + (1 - \eta_{ij})(\boldsymbol{v}_i^j(k) + \boldsymbol{v}_j^i(k)) + \eta_{ij}(\hat{\boldsymbol{v}}_i^j(k) + \hat{\boldsymbol{v}}_j^i(k)) \\ &= \tilde{\boldsymbol{\lambda}}_i^j(k) + \eta_i^j(\hat{\boldsymbol{v}}_i^j(k) + \hat{\boldsymbol{v}}_j^i(k)), \end{aligned}$$

where the second equality is obtained by substituting $\boldsymbol{\lambda}_i^j(k+1)$ with the expression in (3.11) and the fourth equality is obtained by subtituting $\boldsymbol{v}_i^j(k+1)$ and $\boldsymbol{v}_j^i(k+1)$ with the expression in (3.10) and by using the fact that $\eta_i^j = \eta_j^i = \eta_{ij}$. Thus, it follows from the preceding expression that

$$\begin{aligned}\|\tilde{\boldsymbol{\lambda}}_i^j(k+1)-\boldsymbol{\lambda}_i^{j\star}\|_2^2 &= \|\tilde{\boldsymbol{\lambda}}_i^j(k)+\eta_i^j(\hat{\boldsymbol{v}}_i^j(k)+\hat{\boldsymbol{v}}_j^i(k))-\boldsymbol{\lambda}_i^{j\star}\|_2^2\\ &= \|\tilde{\boldsymbol{\lambda}}_i^j(k)-\boldsymbol{\lambda}_i^{j\star}\|_2^2+\|\eta_i^j(\hat{\boldsymbol{v}}_i^j(k)+\hat{\boldsymbol{v}}_j^i(k))\|_2^2\\ &\quad+2\eta_i^j\langle\tilde{\boldsymbol{\lambda}}_i^j(k)-\boldsymbol{\lambda}_i^{j\star},\hat{\boldsymbol{v}}_i^j(k)+\hat{\boldsymbol{v}}_j^i(k)\rangle.\end{aligned} \tag{3.21}$$

Moreover, it holds that

$$\begin{aligned}\|\boldsymbol{v}_i^j(k+1)-\boldsymbol{v}_i^{j\star}\|_2^2 &= \|\boldsymbol{v}_i^j(k)+\eta_i^j(\hat{\boldsymbol{v}}_i^j-\boldsymbol{v}_i^j(k))-\boldsymbol{v}_i^{j\star}\|_2^2\\ &= \|\boldsymbol{v}_i^j(k)-\boldsymbol{v}_i^{j\star}\|_2^2+\|\eta_i^j(\hat{\boldsymbol{v}}_i^j-\boldsymbol{v}_i^j(k))\|_2^2\\ &\quad+2\eta_i^j\langle\hat{\boldsymbol{v}}_i^j-\boldsymbol{v}_i^j(k),\boldsymbol{v}_i^j(k)-\boldsymbol{v}_i^{j\star}\rangle.\end{aligned} \tag{3.22}$$

From (3.21) and (3.22), the term $V_1(k+1)-V_1(k)$ can be expressed, as follows:

$$\begin{aligned}V_1(k+1)-V_1(k) &= \|\hat{\boldsymbol{v}}(k)-\boldsymbol{v}(k)\|_{H^{-1}}^2+\sum_{i\in\mathcal{N}}\sum_{j\in\mathcal{N}_i}\frac{\eta_i^j}{2}\|(\hat{\boldsymbol{v}}_i^j(k)+\hat{\boldsymbol{v}}_j^i(k))\|_2^2\\ &\quad+\sum_{i\in\mathcal{N}}\sum_{j\in\mathcal{N}_i}\langle\tilde{\boldsymbol{\lambda}}_i^j(k)-\boldsymbol{\lambda}_i^{j\star},\hat{\boldsymbol{v}}_i^j(k)+\hat{\boldsymbol{v}}_j^i(k)\rangle\\ &\quad+\sum_{i\in\mathcal{N}}\sum_{j\in\mathcal{N}_i}2\langle\hat{\boldsymbol{v}}_i^j-\boldsymbol{v}_i^j(k),\boldsymbol{v}_i^j(k)-\boldsymbol{v}_i^{j\star}\rangle\\ &\le \|\hat{\boldsymbol{v}}(k)-\boldsymbol{v}(k)\|_{H^{-1}}^2+\sum_{i\in\mathcal{N}}\sum_{j\in\mathcal{N}_i}\frac{\eta_i^j}{2}\|(\hat{\boldsymbol{v}}_i^j(k)+\hat{\boldsymbol{v}}_j^i(k))\|_2^2\\ &\quad-\frac{3}{2}\|\hat{\boldsymbol{v}}(k)-\boldsymbol{v}(k)\|_2^2-\sum_{i\in\mathcal{N}}\sigma_i\|\boldsymbol{u}_i(k+1)-\boldsymbol{u}_i^\star\|_2^2\\ &\quad-\sum_{i\in\mathcal{N}}\sum_{j\in\mathcal{N}_i}\frac{2\eta_i^j-(2\eta_i^j)^2}{2}\|\hat{\boldsymbol{v}}_i^j(k)+\hat{\boldsymbol{v}}_j^i(k)\|_2^2,\end{aligned}$$

where the inequality is obtained by using (3.18), which provides the estimate of the last two term on the right-hand side of the first equality. Thus, the inequality (3.20) follows. Moreover, based on (3.20), $V_1(k)$ is monotonically non-increasing, i.e., $V_1(k+1)-V_1(k)\le 0$ if $\eta_i^j=\eta_j^i=\eta_{ij}\in(0,\frac{1}{4})$.

Finally, the convergence properties of Algorithm 3.1 are stated as follows.

Theorem 3.1 *Let Assumptions 2.1–2.3 and 3.1 hold. Furthermore, let the sequence* $\{(\boldsymbol{u}(k),\boldsymbol{v}(k),\hat{\boldsymbol{v}}(k),\boldsymbol{\lambda}(k))\}$ *be generated by Algorithm 3.1. If* $\eta_i^j=\eta_j^i=\eta_{ij}\in\left(0,\frac{1}{4}\right)$, *for all* $j\in\mathcal{N}$ *and* $i\in\mathcal{N}$, *then,*

a. *(Feasibility)* $\lim_{k\to\infty}\|\boldsymbol{v}_i^j(k)+\boldsymbol{v}_j^i(k)\|_2^2=0$, *for all* $j\in\mathcal{N}_i$ *and* $i\in\mathcal{N}$,
b. *(Primal and dual variable convergence) There exists a saddle point of* $L_1(\boldsymbol{u},\boldsymbol{v},\boldsymbol{\lambda})$ *(see (3.3)), denoted by* $(\boldsymbol{u}^\star,\boldsymbol{v}^\star,\boldsymbol{\lambda}^\star)$, *such that* $\lim_{k\to\infty}\boldsymbol{u}(k)=\boldsymbol{u}^\star$, $\lim_{k\to\infty}\boldsymbol{v}(k)=\boldsymbol{v}^\star$, *and* $\lim_{k\to\infty}\boldsymbol{\lambda}(k)=\boldsymbol{\lambda}^\star$. □

Proof Recall the function $V_1(k)$ defined in (3.19) and the inequality (3.20) in Lemma 3.3. By rearranging and iterating (3.20), for $\ell = 0, \ldots, k$, it follows that

$$\begin{aligned}
&\sum_{\ell=0}^{k}\sum_{i\in\mathcal{N}} \sigma_i \|\boldsymbol{u}_i(\ell+1) - \boldsymbol{u}_i^\star\|_2^2 \\
&+ \sum_{\ell=0}^{k}\sum_{i\in\mathcal{N}}\sum_{j\in\mathcal{N}_i} \left(\frac{3}{2} - \eta_{ij}\right) \|\hat{\boldsymbol{v}}_i^j(\ell) - \boldsymbol{v}_i^j(\ell)\|_2^2 \\
&+ \sum_{\ell=0}^{k}\sum_{i\in\mathcal{N}}\sum_{j\in\mathcal{N}_i} \frac{\eta_{ij} - (2\eta_{ij})^2}{2} \|\hat{\boldsymbol{v}}_i^j(\ell) + \hat{\boldsymbol{v}}_j^i(\ell)\|_2^2 \\
&\leq \sum_{\ell=0}^{k} (V_1(\ell) - V_1(\ell+1)) = V_1(0) - V_1(k+1) \leq V_1(0),
\end{aligned}$$

where the last inequality is obtained by dropping the non-positive term $-V_1(k+1)$. The above inequalities imply that each each quadratic term on the left-hand side of the first inequality is summable and converges to 0 as k goes to infinity, i.e., for each $i \in \mathcal{N}$,

$$\lim_{k\to\infty} \|\boldsymbol{u}_i(k) - \boldsymbol{u}_i^\star\|_2^2 = 0, \tag{3.23}$$

$$\lim_{k\to\infty} \|\boldsymbol{v}_i^j(k) - \hat{\boldsymbol{v}}_i^j(k)\|_2^2 = 0, \ \forall j \in \mathcal{N}_i, \tag{3.24}$$

$$\lim_{k\to\infty} \|\hat{\boldsymbol{v}}_i^j(k) + \hat{\boldsymbol{v}}_j^i(k)\|_2^2 = 0, \ \forall j \in \mathcal{N}_i. \tag{3.25}$$

Based on (3.24) and (3.25), it follows that

$$\lim_{k\to\infty} \|\boldsymbol{v}_i^j(k) + \boldsymbol{v}_j^i(k)\|_2^2 = 0, \ \forall j \in \mathcal{N}_i, \ \forall i \in \mathcal{N}. \tag{3.26}$$

Based on (3.20), the sequences $\{\|\boldsymbol{v}(k) - \boldsymbol{v}^\star\|_{\tilde{H}}^2\}$ and $\{\|\tilde{\boldsymbol{\lambda}}(k) - \boldsymbol{\lambda}^\star\|_{\tilde{H}}^2\}$ are bounded. Thus, there exist accumulation points of the sequences $\{\boldsymbol{v}(k)\}$ and $\{\tilde{\boldsymbol{\lambda}}(k)\}$. Furthermore, due to the boundedness of $\{\tilde{\boldsymbol{\lambda}}(k)\}$ and (3.26), $\{\boldsymbol{\lambda}(k)\}$ is also bounded and has accumulation points. Let the subsequence $\{(\boldsymbol{v}(k_\ell), \boldsymbol{\lambda}(k_\ell))\}$ be convergent and denote its limit point by $(\boldsymbol{v}^{\mathrm{a}}, \boldsymbol{\lambda}^{\mathrm{a}})$.

The next step of the proof is to show that $(\boldsymbol{u}^\star, \boldsymbol{v}^{\mathrm{a}}, \boldsymbol{\lambda}^{\mathrm{a}})$ is a saddle point of $L_1(\boldsymbol{u}, \boldsymbol{v}, \boldsymbol{\lambda})$, i.e., $(\boldsymbol{u}^\star, \boldsymbol{v}^{\mathrm{a}}, \boldsymbol{\lambda}^{\mathrm{a}})$ satisfies the inequalities in (3.6). Based on (3.26), $\boldsymbol{v}_i^{j\mathrm{a}} + \boldsymbol{v}_j^{i\mathrm{a}} = \lim_{\ell\to\infty}(\boldsymbol{v}_i^j(k_\ell) + \boldsymbol{v}_j^i(k_\ell)) = 0$, for all $j \in \mathcal{N}_i$ and $i \in \mathcal{N}$. Thus, it is obtained that, for any $\boldsymbol{\lambda} \in \mathbb{R}^{\sum_{i\in\mathcal{N}} h|\mathcal{N}_i|}$, $L_1(\boldsymbol{u}^\star, \boldsymbol{v}^{\mathrm{a}}, \boldsymbol{\lambda}) = L_1(\boldsymbol{u}^\star, \boldsymbol{v}^{\mathrm{a}}, \boldsymbol{\lambda}^{\mathrm{a}})$, satisfying the first inequality in (3.6). It remains to show the second inequality in (3.6). Based on (3.24), for all $i \in \mathcal{N}$, it follows that $\lim_{\ell\to\infty} \hat{\boldsymbol{v}}(k_\ell) = \boldsymbol{v}^{\mathrm{a}}$. Now, consider the update step (3.9), for all $i \in \mathcal{N}$, i.e.,

$$(\boldsymbol{u}(k+1), \hat{\boldsymbol{v}}(k)) = \arg \min_{(\boldsymbol{u}_i, \boldsymbol{v}_i) \in \mathcal{L}_i, i \in \mathcal{N}} \sum_{i \in \mathcal{N}} \Big(f_i^{\mathrm{l}}(\boldsymbol{u}_i) + f_i^{\mathrm{c}}(\boldsymbol{v}_i) + \sum_{j \in \mathcal{N}_i} \Big(\langle \boldsymbol{\lambda}_i^j(k) + \boldsymbol{\lambda}_j^i(k), \boldsymbol{v}_i^j \rangle + \|\boldsymbol{v}_i^j + \boldsymbol{v}_j^i(k)\|_2^2 \Big) \Big).$$

By substituting k with k_ℓ, it follows that $\lim_{\ell \to \infty} (\boldsymbol{u}(k_\ell + 1), \hat{\boldsymbol{v}}(k_\ell)) = (\boldsymbol{u}^\star, \boldsymbol{v}^{\mathrm{a}})$, based on (5.12), and it also holds that

$$\begin{aligned}
(\boldsymbol{u}^\star, \boldsymbol{v}^{\mathrm{a}}) &= \lim_{\ell \to \infty} \arg \min_{(\boldsymbol{u}_i, \boldsymbol{v}_i) \in \mathcal{L}_i, i \in \mathcal{N}} \sum_{i \in \mathcal{N}} \Big(f_i^{\mathrm{l}}(\boldsymbol{u}_i) + f_i^{\mathrm{c}}(\boldsymbol{v}_i) \\
&\quad + \sum_{j \in \mathcal{N}_i} \Big(\langle \boldsymbol{\lambda}_i^j(k_\ell) + \boldsymbol{\lambda}_j^i(k_\ell), \boldsymbol{v}_i^j \rangle + \|\boldsymbol{v}_i^j + \boldsymbol{v}_j^i(k_\ell)\|_2^2 \Big) \Big) \\
&= \arg \min_{(\boldsymbol{u}_i, \boldsymbol{v}_i) \in \mathcal{L}_i, i \in \mathcal{N}} \sum_{i \in \mathcal{N}} \Big(f_i^{\mathrm{l}}(\boldsymbol{u}_i) + f_i^{\mathrm{c}}(\boldsymbol{v}_i) + \sum_{j \in \mathcal{N}_i} \Big(\langle \boldsymbol{\lambda}_i^{j\mathrm{a}} + \boldsymbol{\lambda}_j^{i\mathrm{a}}, \boldsymbol{v}_i^j \rangle + \|\boldsymbol{v}_i^j + \boldsymbol{v}_j^{i\mathrm{a}}\|_2^2 \Big) \Big) \\
&= \arg \min_{(\boldsymbol{u}_i, \boldsymbol{v}_i) \in \mathcal{L}_i, i \in \mathcal{N}} \sum_{i \in \mathcal{N}} \Big(f_i^{\mathrm{l}}(\boldsymbol{u}_i) + f_i^{\mathrm{c}}(\boldsymbol{v}_i) + \sum_{j \in \mathcal{N}_i} \langle \boldsymbol{\lambda}_i^{j\mathrm{a}}, \boldsymbol{v}_i^j + \boldsymbol{v}_j^i \rangle \Big),
\end{aligned} \tag{3.27}$$

where the last equality holds by removing the term $\sum_{i \in \mathcal{N}} \sum_{j \in \mathcal{N}_i} \|\boldsymbol{v}_i^j + \boldsymbol{v}_j^{i\mathrm{a}}\|_2^2$ since this term is zero at $(\boldsymbol{u}^\star, \boldsymbol{v}^{\mathrm{a}})$ due to the fact that $\boldsymbol{v}_i^{j\mathrm{a}} + \boldsymbol{v}_j^{i\mathrm{a}} = 0$, for all $j \in \mathcal{N}_i$ and $i \in \mathcal{N}$. Additionally, $\boldsymbol{v}^{\mathrm{a}}$ is an attainer of $\min_{\boldsymbol{v}} \sum_{i \in \mathcal{N}} \sum_{j \in \mathcal{N}_i} \|\boldsymbol{v}_i^j + \boldsymbol{v}_j^i\|_2^2$. Therefore, the pair $(\boldsymbol{u}^\star, \boldsymbol{v}^{\mathrm{a}})$ also minimizes $L_1(\boldsymbol{u}, \boldsymbol{v}, \boldsymbol{\lambda}^{\mathrm{a}})$, i.e.,

$$(\boldsymbol{u}^\star, \boldsymbol{v}^{\mathrm{a}}) \in \arg \min_{(\boldsymbol{u}_i, \boldsymbol{v}_i) \in \mathcal{L}_i, i \in \mathcal{N}} \sum_{i \in \mathcal{N}} \Big(f_i^{\mathrm{l}}(\boldsymbol{u}_i) + f_i^{\mathrm{c}}(\boldsymbol{v}_i) + \sum_{j \in \mathcal{N}_i} \Big(\langle \boldsymbol{\lambda}_i^{j\mathrm{a}}, \boldsymbol{v}_i^j + \boldsymbol{v}_j^i \rangle + \|\boldsymbol{v}_i^j + \boldsymbol{v}_j^i\|_2^2 \Big) \Big),$$

where the cost function in the minimization is obtained by adding the quadratic term $\sum_{i \in \mathcal{N}} \sum_{j \in \mathcal{N}_i} \|\boldsymbol{v}_i^j + \boldsymbol{v}_j^i\|_2^2$ to the cost function on the right-hand side of the equality in (3.27). Hence, the preceding relation implies the second inequality in (3.6). Thus, $(\boldsymbol{u}^\star, \boldsymbol{v}^{\mathrm{a}}, \boldsymbol{\lambda}^{\mathrm{a}})$ is a saddle point of $L_1(\boldsymbol{u}, \boldsymbol{v}, \boldsymbol{\lambda})$. Finally, set $\boldsymbol{v}^\star = \boldsymbol{v}^{\mathrm{a}}$ and $\boldsymbol{\lambda}^\star = \boldsymbol{\lambda}^{\mathrm{a}}$ in $V_1(k)$ (see (3.19)). Since the subsequence of $V_1(k_\ell)$ converges to 0 and due to the monotonicity of $V_1(k)$, the entire sequence converges to $(\boldsymbol{v}^{\mathrm{a}}, \boldsymbol{\lambda}^{\mathrm{a}})$.

Remark 3.1 If the dual variables $\boldsymbol{\lambda}_i(0)$, for all $i \in \mathcal{N}$, are initialized such that $\boldsymbol{\lambda}_{i0}^j = \boldsymbol{\lambda}_{j0}^i$, then, it follows that $\boldsymbol{\lambda}_i^j(k) = \boldsymbol{\lambda}_j^i(k)$, for all $k \in \mathbb{Z}_{\geq 0}$, since $\eta_i^j = \eta_j^i = \eta_{ij}$. In this setup, the second round of communication (Step 5) in Algorithm 3.1 is not necessary and each agent $p \in \mathcal{P}$ knows $\boldsymbol{\lambda}_j^i(k+1)$, for all $j \in \mathcal{N}_i \backslash \mathcal{M}_p$ and $i \in \mathcal{M}_p$, at each step k. □

3.3 Distributed ADMM Algorithm

The second distributed algorithm presented in this chapter is based on the ADMM approach [2, 18]. Similar to Algorithm 3.1, it is an iterative algorithm that solves the dual of an augmented problem associated with Problem (3.1). In this regard, at each iteration, primal and dual variables are updated. However, in the ADMM-based algorithm, the information from the neighbors is not necessary when performing the local optimization step to compute primal variables, as can be seen later. Note that Problem (3.1) has a special structure due to the edge-based coupling constraints and the partition of the decision variables into coupled and local constraints. In the derivation of the algorithm, this structure is exploited, similarly as in [11].

3.3.1 Algorithm Design

First, an auxiliary variables $\boldsymbol{y}_i^j \in \mathbb{R}^h$, for each $j \in \mathcal{N}_i$ and $i \in \mathcal{N}$, is introduced. Note that for convenience, these variables are compactly written as $\boldsymbol{y}_i = \mathrm{col}(\{\boldsymbol{y}_i^j\}_{j\in\mathcal{N}_i})$ and $\boldsymbol{y} = \mathrm{col}(\{\boldsymbol{y}_i\}_{i\in\mathcal{N}})$. Then, Problem (3.1) is reformulated and augmented as follows:

$$\underset{\{(\boldsymbol{u}_i,\boldsymbol{v}_i)\in\mathcal{L}_i,\boldsymbol{y}_i\in\mathbb{R}^{h|\mathcal{N}_i|}\}_{i\in\mathcal{N}}}{\text{minimize}} \sum_{p\in\mathcal{P}}\sum_{i\in\mathcal{M}_p}\left(f_i^{\mathrm{l}}(\boldsymbol{u}_i) + f_i^{\mathrm{c}}(\boldsymbol{v}_i) + \sum_{j\in\mathcal{N}_i}\frac{1}{2}\|\boldsymbol{v}_i^j - \boldsymbol{y}_i^j\|_2^2\right) \tag{3.28a}$$

$$\text{s.t. } \boldsymbol{v}_i^j - \boldsymbol{y}_i^j = 0, \quad \forall j \in \mathcal{N}_i,\ \forall i \in \mathcal{M}_p,\ \forall p \in \mathcal{P}, \tag{3.28b}$$

$$\boldsymbol{y}_i^j + \boldsymbol{y}_j^i = 0, \quad \forall j \in \mathcal{N}_i,\ \forall i \in \mathcal{M}_p,\ \forall p \in \mathcal{P}. \tag{3.28c}$$

The idea behind the introduction of the auxiliary variables and the addition of the constraints in (3.28b) is to construct a separable cost function of the augmented problem (3.28). As can be seen, not only the cost function (3.28a) but also (3.28b) can be suitably decomposed to the subsystems. Meanwhile, the coupling constraints between neighboring subsystems appear in some of the constraints in (3.28c), i.e., $\boldsymbol{y}_i^j + \boldsymbol{y}_j^i = 0$, for each $j \in \mathcal{N}_i \backslash \mathcal{M}_p$, $i \in \mathcal{M}_p$, and $p \in \mathcal{P}$.

Then, denote the Lagrange multipliers associated with the constraints in (3.28b) by $\boldsymbol{\mu}_i^j \in \mathbb{R}^{h|\mathcal{N}_i|}$, for all $j \in \mathcal{N}_i$ and $i \in \mathcal{N}$. In this regard, the Lagrangian function, which is associated with Problem (3.28) and denoted by $L_2(\boldsymbol{u}, \boldsymbol{v}, \boldsymbol{y}, \boldsymbol{\mu})$, is defined as follows:

$$L_2(\boldsymbol{u}, \boldsymbol{v}, \boldsymbol{y}, \boldsymbol{\mu}) = \sum_{p\in\mathcal{P}}\sum_{i\in\mathcal{M}_p}\left(f_i^{\mathrm{l}}(\boldsymbol{u}_i) + f_i^{\mathrm{c}}(\boldsymbol{v}_i) + \sum_{j\in\mathcal{N}_i}\left(\langle\boldsymbol{\mu}_i^j, \boldsymbol{v}_i^j - \boldsymbol{y}_i^j\rangle + \frac{1}{2}\|\boldsymbol{v}_i^j - \boldsymbol{y}_i^j\|_2^2\right)\right), \tag{3.29}$$

where $\boldsymbol{\mu} = \operatorname{col}(\{\boldsymbol{\mu}_i\}_{i\in\mathcal{N}})$, $\boldsymbol{\mu}_i = \operatorname{col}(\{\boldsymbol{\mu}_i^j\}_{j\in\mathcal{N}_i})$. Therefore, the dual function associated with (3.28), denoted by $g_2(\boldsymbol{\mu})$, is stated, as follows:

$$\begin{aligned} g_2(\boldsymbol{\mu}) = & \min_{\{(\boldsymbol{u}_i,\boldsymbol{v}_i)\in\mathcal{L}_i,\boldsymbol{y}_i\in\mathbb{R}^{h|\mathcal{N}_i|}\}_{i\in\mathcal{N}}} L_2(\boldsymbol{u},\boldsymbol{v},\boldsymbol{y},\boldsymbol{\mu}) \\ & \text{s.t. } \boldsymbol{y}_i^j + \boldsymbol{y}_j^i = 0, \quad \forall j \in \mathcal{N}_i, \ \forall i \in \mathcal{M}_p, \ \forall p \in \mathcal{P}, \end{aligned} \tag{3.30}$$

and the dual problem associated with (3.28) is as follows:

$$\underset{\boldsymbol{\mu}}{\text{maximize}}\ g_2(\boldsymbol{\mu}).$$

Note that the strong duality holds and the set of dual optimal points is non-empty, for the same reason discussed in Sect. 3.2.1. Therefore, there exists a saddle point of the Lagrangian function $L_2(\boldsymbol{u},\boldsymbol{v},\boldsymbol{y},\boldsymbol{\mu})$, i.e., a point $(\boldsymbol{u}^\star,\boldsymbol{v}^\star,\boldsymbol{y}^\star,\boldsymbol{\mu}^\star) \in \prod_{i\in\mathcal{N}} \mathcal{L}_i \times \mathbb{R}^{2\sum_{i\in\mathcal{N}} h|\mathcal{N}_i|}$ such that for any $(\boldsymbol{u},\boldsymbol{v}) \in \prod_{i\in\mathcal{N}} \mathcal{L}_i$ and $(\boldsymbol{y},\boldsymbol{\mu}) \in \mathbb{R}^{2\sum_{i\in\mathcal{N}} h|\mathcal{N}_i|}$, it holds that

$$L_2(\boldsymbol{u}^\star,\boldsymbol{v}^\star,\boldsymbol{y}^\star,\boldsymbol{\mu}) \le L_2(\boldsymbol{u}^\star,\boldsymbol{v}^\star,\boldsymbol{y}^\star,\boldsymbol{\mu}^\star) \le L_2(\boldsymbol{u},\boldsymbol{v},\boldsymbol{y},\boldsymbol{\mu}^\star), \tag{3.31}$$

implying that $(\boldsymbol{u}^\star,\boldsymbol{v}^\star,\boldsymbol{y}^\star)$ is a solution to Problem (3.28), while $\boldsymbol{\mu}^\star$ is a dual optimal solution.

As in Algorithm 3.1, the ADMM-based algorithm consists of computing the attainer of local minimization problems derived from the dual function to update the primal variables and perform a gradient ascent to update the dual variables. Different from Algorithm 3.1, here the dual function is not approximated. Instead, the primal variables are computed by alternately solving the subproblems obtained from the minimization in (3.30). In particular, the ADMM steps at iteration k are given as follows.

1. Update the primal variables $\boldsymbol{u}(k+1)$ and $\boldsymbol{v}(k+1)$, as follows:

$$(\boldsymbol{u}(k+1),\boldsymbol{v}(k+1)) = \arg \min_{\{(\boldsymbol{u}_i,\boldsymbol{v}_i)\in\mathcal{L}_i\}_{i\in\mathcal{N}}} L_2((\boldsymbol{u},\boldsymbol{v},\boldsymbol{y}(k),\boldsymbol{\mu}(k))). \tag{3.32}$$

2. Update the auxiliary variables $\boldsymbol{y}(k+1)$, as follows:

$$\begin{aligned} \boldsymbol{y}(k+1) = \arg & \min_{\{\boldsymbol{y}_i\in\mathbb{R}^{h|\mathcal{N}_i|}\}_{i\in\mathcal{N}}} L_2(\boldsymbol{u}(k+1),\boldsymbol{v}(k+1),\boldsymbol{y},\boldsymbol{\mu}(k)) \\ & \text{s.t. } \boldsymbol{y}_i^j + \boldsymbol{y}_j^i = 0, \quad \forall j \in \mathcal{N}_i, \ \forall i \in \mathcal{M}_p, \ \forall p \in \mathcal{P}. \end{aligned} \tag{3.33}$$

3. Update the dual variables $\boldsymbol{\mu}(k+1)$, as follows:

$$\boldsymbol{\mu}_i^j(k+1) = \boldsymbol{\mu}_i^j(k) + \eta_i^j\left(\boldsymbol{v}_i^j(k+1) - \boldsymbol{y}_i^j(k+1)\right), \quad \forall j \in \mathcal{N}_i, \ \forall i \in \mathcal{N}, \tag{3.34}$$

where $\eta_i^j \in \mathbb{R}_{\ge 0}$, for each $j \in \mathcal{N}_i$ and $i \in \mathcal{N}$, is the step size to update $\boldsymbol{\mu}_i^j(k+1)$.

Due to the structure of $L_2(\boldsymbol{u}, \boldsymbol{v}, \boldsymbol{y}, \boldsymbol{\mu})$, the minimization in (3.32) is decomposable. Moreover, the dual variable update in (3.34) can also be carried in a non-centralized manner. The remaining task is to decompose the auxiliary variable update in (3.33), which is not trivially decomposable due to the existence of coupling equality constraints. However, the analytical solution to the minimization (3.33) can be computed. It is observed that the minimization in (3.33) is a quadratic problem in the following form:

$$\begin{aligned} &\underset{\{\boldsymbol{y}_i \in \mathbb{R}^{h|\mathcal{N}_i|}\}_{i\in\mathcal{N}}}{\text{minimize}} \sum_{i\in\mathcal{N}} \sum_{j\in\mathcal{N}_i} \left(-\langle \boldsymbol{\mu}_i^j(k), \boldsymbol{y}_i^j \rangle + \frac{1}{2} \|\boldsymbol{v}_i^j(k+1) - \boldsymbol{y}_i^j\|_2^2 \right) \\ &\text{s.t. } \boldsymbol{y}_i^j + \boldsymbol{y}_j^i = 0, \quad \forall j \in \mathcal{N}_i, \ \forall i \in \mathcal{M}_p, \ \forall p \in \mathcal{P}. \end{aligned} \tag{3.35}$$

Since the cost function is strongly convex, the solution is unique. Moreover, the cost function can be written as

$$\sum_{\{i,j\}\in\mathcal{E}} -\langle \boldsymbol{\mu}_i^j(k), \boldsymbol{y}_i^j \rangle - \langle \boldsymbol{\mu}_j^i(k), \boldsymbol{y}_j^i \rangle + \frac{1}{2}\|\boldsymbol{v}_i^j(k+1) - \boldsymbol{y}_i^j\|_2^2 + \frac{1}{2}\|\boldsymbol{v}_j^i(k+1) - \boldsymbol{y}_j^i\|_2^2.$$

By imposing the equality constraints in (3.28c), i.e., $\boldsymbol{y}_j^i = -\boldsymbol{y}_i^j$, the problem in (3.35) is equivalent to

$$\sum_{\{i,j\}\in\mathcal{E}} \underset{\boldsymbol{y}_i^j}{\text{minimize}} \ - \langle \boldsymbol{\mu}_i^j(k), \boldsymbol{y}_i^j \rangle + \langle \boldsymbol{\mu}_j^i(k), \boldsymbol{y}_i^j \rangle + \frac{1}{2}\|\boldsymbol{v}_i^j(k+1) - \boldsymbol{y}_i^j\|_2^2 + \frac{1}{2}\|\boldsymbol{v}_j^i(k+1) + \boldsymbol{y}_i^j\|_2^2.$$

Denote by $\boldsymbol{y}_i^j(k+1)$ the solution of the above minimization, for each $\{i, j\} \in \mathcal{E}$. Thus, the optimality condition of this minimization problem is as follows:

$$0 = -\boldsymbol{\mu}_i^j(k) + \boldsymbol{\mu}_j^i(k) - \left(\boldsymbol{v}_i^j(k+1) - \boldsymbol{y}_i^j(k+1)\right) + \left(\boldsymbol{v}_j^i(k+1) + \boldsymbol{y}_i^j(k+1)\right),$$

implying that

$$\boldsymbol{y}_i^j(k+1) = \frac{1}{2}\left(\boldsymbol{\mu}_i^j(k) - \boldsymbol{\mu}_j^i(k) + \boldsymbol{v}_i^j(k+1) - \boldsymbol{v}_j^i(k+1)\right), \ \forall j \in \mathcal{N}_i, \forall i \in \mathcal{N}.$$

It turns out that $\boldsymbol{y}_i^j(k+1)$, for each node $i \in \mathcal{N}$ can be computed using the information of its neighboring nodes $j \in \mathcal{N}_i$, implying the decomposability of this step. Thus, the distributed ADMM-based algorithm is stated in Algorithm 3.2. Note that the step to update $(\boldsymbol{u}_i(k+1), \boldsymbol{v}_i(k+1))$, in (3.36) is obtained by decomposing (3.32). Then, each agent must exchange some of its coupled variables in order to update $\boldsymbol{y}_i(k+1)$, for all nodes that belong to that agent, as in (3.37). Finally, the dual variable updates, shown in (3.38), only require local information that each agent has acquired from the previous steps.

Remark 3.2 It is assumed that the cost function in Problem (3.1) is dual-friendly [23], i.e., the explicit solution to (3.36) (and also (3.9)) is available. The optimization

Algorithm 3.2 Distributed ADMM-based method

Initialization: For each $i \in \mathcal{N}$, $\boldsymbol{\mu}_i(0) = \boldsymbol{\mu}_{i0} \in \mathbb{R}^{|\mathcal{N}_i|h}$, and $\boldsymbol{y}_i(0) \in \mathbb{R}^{h|\mathcal{N}_i|}$.
Iteration: For each agent $p \in \mathcal{P}$,

1. Update $\boldsymbol{u}_i(k+1)$ and $\boldsymbol{v}_i(k+1)$, for all $i \in \mathcal{M}_p$, according to

$$\begin{aligned}\{(\boldsymbol{u}_i(k+1), \boldsymbol{v}_i(k+1))\}_{i \in \mathcal{M}_p} = \arg \min_{\{(\boldsymbol{u}_i,\boldsymbol{v}_i) \in \mathcal{L}_i\}_{i \in \mathcal{M}_p}} \sum_{i \in \mathcal{M}_p} \Big(f_i^{\mathrm{l}}(\boldsymbol{u}_i) + f_i^{\mathrm{c}}(\boldsymbol{v}_i) \\ + \sum_{j \in \mathcal{N}_i} \langle \boldsymbol{\mu}_i^j(k), \boldsymbol{v}_i^j \rangle + \frac{1}{2} \|\boldsymbol{v}_i^j - \boldsymbol{y}_i^j(k)\|_2^2 \Big).\end{aligned} \tag{3.36}$$

2. For each $j \in \mathcal{N}_i \backslash \mathcal{M}_p$ and $i \in \mathcal{M}_p$, send $\boldsymbol{v}_i^j(k+1)$ and $\boldsymbol{\mu}_i^j(k)$ to agent $\phi(j, \cdot) \in \mathcal{N}_p^{\mathrm{s}}$ and receive $\boldsymbol{v}_j^i(k+1)$ and $\boldsymbol{\mu}_j^i(k)$ from agent $\phi(j, \cdot) \in \mathcal{N}_p^{\mathrm{s}}$.
3. Update $\boldsymbol{y}_i^j(k+1)$, for all $j \in \mathcal{N}_i$ and $i \in \mathcal{M}_p$, as follows:

$$\boldsymbol{y}_i^j(k+1) = \frac{1}{2} \Big(\boldsymbol{\mu}_i^j(k) - \boldsymbol{\mu}_j^i(k) + \boldsymbol{v}_i^j(k+1) - \boldsymbol{v}_j^i(k+1) \Big). \tag{3.37}$$

4. Update the dual variables $\boldsymbol{\mu}_i^j(k+1)$, for all $j \in \mathcal{N}_i$ and $i \in \mathcal{M}_p$, according to

$$\boldsymbol{\mu}_i^j(k+1) = \boldsymbol{\mu}_i^j(k) + \eta_i^j \Big(\boldsymbol{v}_i^j(k+1) - \boldsymbol{y}_i^j(k+1) \Big). \tag{3.38}$$

problems in (3.36) (and (3.9)) are strongly convex, and efficient numerical methods to compute the solution are available [24, 25].

Remark 3.3 In order to initialize $\boldsymbol{y}_i^j(0)$, for all $j \in \mathcal{N}_i \backslash \mathcal{M}_p$ and $i \in \mathcal{M}_p$, agent p must agree with its neighbor $\phi(j, \cdot)$ by means of communication or by setting a common initial point of $\boldsymbol{y}_i^j(0)$, i.e., $\boldsymbol{y}_i^j(0) = \boldsymbol{y}_0 \in \mathbb{R}^h$, for all $j \in \mathcal{N}_i$ and $i \in \mathcal{N}$.

Remark 3.4 Similarly to Algorithm 3.1, Algorithm 3.2 also requires synchronous update and neighbor-to-neighbor communication, as stated in Assumption 3.1.

3.3.2 Convergence Analysis

In this section, the convergence of Algorithm 3.2 is shown. To that end, first denote the residual associated with the relaxed constraint (3.28b) by $\boldsymbol{r}(k) \in \mathbb{R}^{\sum_{i \in \mathcal{N}} h|\mathcal{N}_i|}$, where $\boldsymbol{r}(k) = \mathrm{col}(\{\boldsymbol{r}_i\}_{i \in \mathcal{N}})$, $\boldsymbol{r}_i(k) = \mathrm{col}(\{\boldsymbol{r}_i^j\}_{j \in \mathcal{N}_i})$, and

$$\boldsymbol{r}_i^j(k) = \boldsymbol{v}_i^j(k) - \boldsymbol{y}_i^j(k), \quad \forall j \in \mathcal{N}_i, \ \forall i \in \mathcal{N}. \tag{3.39}$$

The convergence analysis of Algorithm 3.2 follows the same idea as that of Algorithm 3.1. Firstly, a Lyapunov function will be proposed. Then, an upper-bound and the monotonicity of this function will be shown. In order to obtain a useful upper-bound of the Lyapunov function, the optimality conditions of the local optimizations in Algorithm 3.2 and the saddle point inequality in (3.31) are evaluated in Lemma 3.4.

Lemma 3.4 *Let Assumptions 2.1–2.3 hold. Furthermore, let* $(\boldsymbol{u}_i(k+1), \boldsymbol{v}_i(k+1))$*, for all* $i \in \mathcal{M}_p$*, be the attainer of the local optimization in (3.36),* $\boldsymbol{y}_i^j(k+1)$*, for all* $j \in \mathcal{N}_i$ *and* $i \in \mathcal{N}$*, be computed by (3.37), and* $(\boldsymbol{u}^\star, \boldsymbol{v}^\star, \boldsymbol{\mu}^\star)$ *be a saddle point of* $L_1(\boldsymbol{u}, \boldsymbol{v}, \boldsymbol{\mu})$ *as defined in (3.29). Then, it holds that*

$$\begin{aligned}&\langle \boldsymbol{\mu}(k) - \boldsymbol{\mu}^\star, \boldsymbol{r}(k+1)\rangle + \frac{1}{2}\|\boldsymbol{y}(k+1) - \boldsymbol{y}^\star\|_2^2 - \frac{1}{2}\|\boldsymbol{y}(k) - \boldsymbol{y}^\star\|_2^2 \\ &\leq -\frac{1}{2}\|\boldsymbol{r}(k+1)\|_2^2 - \frac{1}{2}\|\boldsymbol{r}(k+1) + \boldsymbol{y}(k+1) - \boldsymbol{y}(k)\|_2^2 - \sum_{i\in\mathcal{N}} \sigma_i \|\boldsymbol{u}_i(k+1) - \boldsymbol{u}_i^\star\|_2^2.\end{aligned} \tag{3.40}$$

Proof Firstly, the optimality condition of the minimization in (3.36) gives

$$\begin{aligned}0 \leq \sum_{i\in\mathcal{M}_p} \Big(&\langle \nabla f_i^{\mathrm{l}}(\boldsymbol{u}_i(k+1)), \boldsymbol{u}_i - \boldsymbol{u}_i(k+1)\rangle + \langle \nabla f_i^{\mathrm{c}}(\boldsymbol{v}_i(k+1)), \boldsymbol{v}_i - \boldsymbol{v}_i(k+1)\rangle \\ &+ \sum_{j\in\mathcal{N}_i} \langle \boldsymbol{\mu}_i^j(k) + \boldsymbol{v}_i^j(k+1) - \boldsymbol{y}_i^j(k), \boldsymbol{v}_i^j - \boldsymbol{v}_i^j(k+1)\rangle \Big),\end{aligned} \tag{3.41}$$

for all $(\boldsymbol{u}_i, \boldsymbol{v}_i) \in \mathcal{L}_i$, where $i \in \mathcal{M}_p$. Moreover, $(\boldsymbol{u}_i(k+1), \boldsymbol{v}_i(k+1)) \in \mathcal{L}_i$, for each $i \in \mathcal{M}_p$.

Secondly, since $\boldsymbol{y}(k+1)$ is computed by (3.37), implying that it is the attainer of (3.35), the optimality condition of the minimization in (3.35) yields

$$0 \leq \sum_{i\in\mathcal{N}} \sum_{j\in\mathcal{N}_i} \langle -\boldsymbol{\mu}_i^j(k) - \boldsymbol{v}_i^j(k+1) + \boldsymbol{y}_i^j(k+1), \boldsymbol{y}_i^j - \boldsymbol{y}_i^j(k+1)\rangle, \tag{3.42}$$

for all $\boldsymbol{y} \in \mathbb{R}^{\sum_{i\in\mathcal{N}} h|\mathcal{N}_i|}$ such that $\boldsymbol{y}_i^j = -\boldsymbol{y}_j^i$. Moreover, $\boldsymbol{y}_i^j(k+1) = -\boldsymbol{y}_j^i(k+1)$, for all $j \in \mathcal{N}_i$ and $\imath \in \mathcal{N}$.

Thirdly, by the second saddle point inequality in (3.31), it holds that

$$\begin{aligned}(\boldsymbol{u}^\star, \boldsymbol{v}^\star, \boldsymbol{y}^\star, \boldsymbol{\mu}^\star) = \arg &\min_{\{(\boldsymbol{u}_i, \boldsymbol{v}_i)\in\mathcal{L}_i, \boldsymbol{y}_i\in\mathbb{R}^{h|\mathcal{N}_i|}\}_{i\in\mathcal{N}}} L_2(\boldsymbol{u}, \boldsymbol{v}, \boldsymbol{y}, \boldsymbol{\mu}) \\ &\text{s.t. } \boldsymbol{y}_i^j + \boldsymbol{y}_j^i = 0, \quad \forall j \in \mathcal{N}_i, \ \forall i \in \mathcal{M}_p, \ \forall p \in \mathcal{P}.\end{aligned} \tag{3.43}$$

Since $(\boldsymbol{u}(k+1), \boldsymbol{v}(k+1), \boldsymbol{y}(k+1)$ is a feasible solution to the minimization in (3.43), i.e., $(\boldsymbol{u}_i(k+1), \boldsymbol{v}_i(k+1)) \in \mathcal{L}_i$, for all $i \in \mathcal{N}$, and $\boldsymbol{y}_i^j(k+1) = -\boldsymbol{y}_j^i(k+1)$, for all $j \in \mathcal{N}_i$ and $\imath \in \mathcal{N}$, the optimality conditions of (3.43) yields

$$
\begin{aligned}
0 \leq \sum_{i \in \mathcal{N}} \Bigg(& \langle \nabla f_i^{1}(\boldsymbol{u}_i^\star), \boldsymbol{u}_i(k+1) - \boldsymbol{u}_i^\star \rangle + \langle \nabla f_i^{c}(\boldsymbol{v}_i^\star), \boldsymbol{v}_i(k+1) - \boldsymbol{v}_i^\star \rangle \\
& + \sum_{j \in \mathcal{N}_i} \Big(\langle \boldsymbol{\mu}_i^{j\star} + \boldsymbol{v}_i^{j\star} - \boldsymbol{y}_i^{j\star}, \boldsymbol{v}_i^j(k+1) - \boldsymbol{v}_i^{j\star} \rangle \\
& \qquad + \langle -\boldsymbol{\mu}_i^{j\star} - \boldsymbol{v}_i^{j\star} + \boldsymbol{y}_i^{j\star}, \boldsymbol{y}_i^j(k+1) - \boldsymbol{y}_i^{j\star} \rangle \Big) \Bigg) \qquad (3.44) \\
= \sum_{i \in \mathcal{N}} \Bigg(& \langle \nabla f_i^{1}(\boldsymbol{u}_i^\star), \boldsymbol{u}_i(k+1) - \boldsymbol{u}_i^\star \rangle + \langle \nabla f_i^{c}(\boldsymbol{v}_i^\star), \boldsymbol{v}_i(k+1) - \boldsymbol{v}_i^\star \rangle \\
& + \sum_{j \in \mathcal{N}_i} \langle \boldsymbol{\mu}_i^{j\star}, \boldsymbol{r}_i^j(k+1) \rangle \Bigg),
\end{aligned}
$$

where the equality is obtained by combining the last two inner products and using the fact that $\boldsymbol{v}_i^{j\star} - \boldsymbol{y}_i^{j\star} = 0$ and $\boldsymbol{r}_i^j(k+1) = \boldsymbol{v}_i^j(k+1) - \boldsymbol{y}_i^j(k+1)$, for all $j \in \mathcal{N}_i$ and $i \in \mathcal{N}$ (see (3.39)). Additionally, $(\boldsymbol{u}_i^\star, \boldsymbol{v}_i^\star) \in \mathcal{L}_i$, for all $i \in \mathcal{N}$ and $\boldsymbol{y}_i^{j\star} + \boldsymbol{y}_j^{i\star} = 0$, for all $j \in \mathcal{N}_i$ and $i \in \mathcal{N}$, implying one can substitute $(\boldsymbol{u}_i, \boldsymbol{v}_i)$ in (3.41) with $(\boldsymbol{u}_i^\star, \boldsymbol{v}_i^\star)$ and $\boldsymbol{y}$ in (3.42) with $\boldsymbol{y}^\star$.

Then, by combining (3.41), for all $p \in \mathcal{P}$, where $(\boldsymbol{u}_i, \boldsymbol{v}_i) = (\boldsymbol{u}_i^\star, \boldsymbol{v}_i^\star)$, for all $i \in \mathcal{M}_p$, (3.42) where $\boldsymbol{y} = \boldsymbol{y}^\star$, and (3.44), it holds that

$$
\begin{aligned}
0 \leq & \sum_{i \in \mathcal{N}} \Bigg(\langle \nabla f_i^{1}(\boldsymbol{u}_i^\star) - f_i^{1}(\boldsymbol{u}_i(k+1)), \boldsymbol{u}_i(k+1) - \boldsymbol{u}_i^\star \rangle + \sum_{j \in \mathcal{N}_i} \langle \boldsymbol{\mu}_i^{j\star}, \boldsymbol{r}_i^j(k+1) \rangle \Bigg) \\
& + \sum_{i \in \mathcal{N}} \langle \nabla f_i^{c}(\boldsymbol{v}_i^\star) - f_i^{c}(\boldsymbol{v}_i(k+1)), \boldsymbol{v}_i(k+1) - \boldsymbol{v}_i^\star \rangle \\
& + \sum_{i \in \mathcal{N}} \sum_{j \in \mathcal{N}_i} \langle \boldsymbol{\mu}_i^j(k) + \boldsymbol{v}_i^j(k+1) - \boldsymbol{y}_i^j(k), \boldsymbol{v}_i^{j\star} - \boldsymbol{v}_i^j(k+1) \rangle \\
& + \sum_{i \in \mathcal{N}} \sum_{j \in \mathcal{N}_i} \langle -\boldsymbol{\mu}_i^j(k) - \boldsymbol{v}_i^j(k+1) + \boldsymbol{y}_i^j(k+1), \boldsymbol{y}_i^{j\star} - \boldsymbol{y}_i^j(k+1) \rangle.
\end{aligned}
\qquad (3.45)
$$

Based on the strong convexity of $f_i^{1}(\boldsymbol{u}_i)$ (see Property 3.2) and the convexity of $f_i^{c}(\boldsymbol{v}_i)$ (see Property 3.1), for all $i \in \mathcal{N}$, it follows from (3.45) that

$$
\begin{aligned}
0 \leq & \sum_{i \in \mathcal{N}} \left(-\sigma_i \|\boldsymbol{u}_i(k+1) - \boldsymbol{u}_i^\star\|_2^2 + \sum_{j \in \mathcal{N}_i} \langle \boldsymbol{\mu}_i^{j\star}, \boldsymbol{r}_i^j(k+1) \rangle \right) \\
& + \sum_{i \in \mathcal{N}} \sum_{j \in \mathcal{N}_i} \langle \boldsymbol{\mu}_i^j(k) + \boldsymbol{v}_i^j(k+1) - \boldsymbol{y}_i^j(k), \boldsymbol{v}_i^{j\star} - \boldsymbol{v}_i^j(k+1) \rangle \\
& + \sum_{i \in \mathcal{N}} \sum_{j \in \mathcal{N}_i} \langle -\boldsymbol{\mu}_i^j(k) - \boldsymbol{v}_i^j(k+1) + \boldsymbol{y}_i^j(k+1), \boldsymbol{y}_i^{j\star} - \boldsymbol{y}_i^j(k+1) \rangle \\
= & \sum_{i \in \mathcal{N}} \left(-\sigma_i \|\boldsymbol{u}_i(k+1) - \boldsymbol{u}_i^\star\|_2^2 + \sum_{j \in \mathcal{N}_i} \langle \boldsymbol{\mu}_i^{j\star} - \boldsymbol{\mu}_i^j(k), \boldsymbol{r}_i^j(k+1) \rangle \right) \\
& + \sum_{i \in \mathcal{N}} \sum_{j \in \mathcal{N}_i} \langle \boldsymbol{v}_i^j(k+1) - \boldsymbol{y}_i^j(k), \boldsymbol{v}_i^{j\star} - \boldsymbol{v}_i^j(k+1) \rangle \\
& + \sum_{i \in \mathcal{N}} \sum_{j \in \mathcal{N}_i} \langle -\boldsymbol{v}_i^j(k+1) + \boldsymbol{y}_i^j(k+1), \boldsymbol{y}_i^{j\star} - \boldsymbol{y}_i^j(k+1) \rangle.
\end{aligned}
\tag{3.46}
$$

The remaining step is to manipulate the last two inner products on the right-hand side of the equality in (3.46), as follows:

$$
\begin{aligned}
& \langle \boldsymbol{v}_i^j(k+1) - \boldsymbol{y}_i^j(k), \boldsymbol{v}_i^{j\star} - \boldsymbol{v}_i^j(k+1) \rangle + \langle -\boldsymbol{v}_i^j(k+1) + \boldsymbol{y}_i^j(k+1), \boldsymbol{y}_i^{j\star} - \boldsymbol{y}_i^j(k+1) \rangle \\
& = \langle \boldsymbol{r}_i^j(k+1) + \boldsymbol{y}_i^j(k+1) - \boldsymbol{y}_i^j(k), \boldsymbol{y}_i^{j\star} - \boldsymbol{r}_i^j(k+1) - \boldsymbol{y}_i^j(k+1) \rangle \\
& \quad + \langle \boldsymbol{r}_i^j(k+1), \boldsymbol{y}_i^j(k+1) - \boldsymbol{y}_i^{j\star} \rangle \\
& = -\|\boldsymbol{r}_i^j(k+1)\|_2^2 + \langle \boldsymbol{r}_i^j(k+1), \boldsymbol{y}_i^{j\star} - \boldsymbol{y}_i^j(k+1) \rangle + \langle \boldsymbol{r}_i^j(k+1), \boldsymbol{y}_i^j(k+1) - \boldsymbol{y}_i^{j\star} \rangle \\
& \quad + \langle \boldsymbol{y}_i^j(k+1) - \boldsymbol{y}_i^j(k), \boldsymbol{y}_i^{j\star} - \boldsymbol{y}_i^j(k+1) \rangle - \langle \boldsymbol{y}_i^j(k+1) - \boldsymbol{y}_i^j(k), \boldsymbol{r}_i^j(k+1) \rangle \\
& = -\|\boldsymbol{r}_i^j(k+1)\|_2^2 + \langle \boldsymbol{y}_i^j(k+1) - \boldsymbol{y}_i^j(k), \boldsymbol{y}_i^{j\star} - \boldsymbol{y}_i^j(k+1) \rangle \\
& \quad - \langle \boldsymbol{y}_i^j(k+1) - \boldsymbol{y}_i^j(k), \boldsymbol{r}_i^j(k+1) \rangle.
\end{aligned}
\tag{3.47}
$$

Furthermore, observe that

$$
\begin{aligned}
& 2\langle \boldsymbol{y}_i^j(k) - \boldsymbol{y}_i^j(k+1), \boldsymbol{y}_i^j(k+1) - \boldsymbol{y}_i^{j\star} \rangle \\
& = -\|\boldsymbol{y}_i^j(k+1) - \boldsymbol{y}_i^{j\star}\|_2^2 - \|\boldsymbol{y}_i^j(k) - \boldsymbol{y}_i^j(k+1)\|_2^2 + \|\boldsymbol{y}_i^j(k) - \boldsymbol{y}_i^{j\star}\|_2^2,
\end{aligned}
\tag{3.48}
$$

$$
\begin{aligned}
-2\langle \boldsymbol{y}_i^j(k+1) - \boldsymbol{y}_i^j(k), \boldsymbol{r}_i^j(k+1) \rangle = & \|\boldsymbol{r}_i^j(k+1)\|_2^2 + \|\boldsymbol{y}_i^j(k+1) - \boldsymbol{y}_i^j(k)\|_2^2 \\
& - \|\boldsymbol{r}_i^j(k+1) + \boldsymbol{y}_i^j(k+1) - \boldsymbol{y}_i^j(k)\|_2^2.
\end{aligned}
\tag{3.49}
$$

By applying the relations (3.48) and (3.49) to (3.47), it holds that

$$\begin{aligned}&\langle \boldsymbol{v}_i^j(k+1) - \boldsymbol{y}_i^j(k), \boldsymbol{v}_i^{j\star} - \boldsymbol{v}_i^j(k+1)\rangle + \langle -\boldsymbol{v}_i^j(k+1) + \boldsymbol{y}_i^j(k+1), \boldsymbol{y}_i^{j\star} - \boldsymbol{y}_i^j(k+1)\rangle \\ &= -\frac{1}{2}\|\boldsymbol{r}_i^j(k+1)\|_2^2 - \frac{1}{2}\|\boldsymbol{y}_i^j(k+1) - \boldsymbol{y}_i^{j\star}\|_2^2 + \frac{1}{2}\|\boldsymbol{y}_i^j(k) - \boldsymbol{y}_i^{j\star}\|_2^2 \\ &\quad -\frac{1}{2}\|\boldsymbol{r}_i^j(k+1) + \boldsymbol{y}_i^j(k+1) - \boldsymbol{y}_i^j(k)\|_2^2. \end{aligned} \tag{3.50}$$

By combining (3.46) and (3.50), the desired inequality follows.

The result obtained in Lemma 3.4 provides an insight of a suitable Lyapunov function that works for Algorithm 3.2. In particular, define the candidate Lyapunov function, denoted by $V_2(k)$, as follows. Recall that $H = \text{blkdiag}(\{H_i\}_{i\in\mathcal{N}})$ and $H_i = \text{blkdiag}(\{(\eta_i^j)^{-1}I_h\}_{j\in\mathcal{N}_i})$, for all $i \in \mathcal{N}$. Then, for any saddle point of $L_2(\boldsymbol{u}, \boldsymbol{v}, \boldsymbol{y}, \boldsymbol{\mu})$, denoted by $(\boldsymbol{u}^\star, \boldsymbol{v}^\star, \boldsymbol{y}^\star, \boldsymbol{\mu})$, let $V_2(k)$ be

$$V_2(k) = \frac{1}{2}\|\boldsymbol{\mu}(k) - \boldsymbol{\mu}^\star\|_H^2 + \frac{1}{2}\|\boldsymbol{y}(k) - \boldsymbol{y}^\star\|_2^2. \tag{3.51}$$

In Lemma 3.5, a sufficient condition such that $V_2(k)$ is monotonically non-increasing and an upper-bound of $V_2(k)$ are provided.

Lemma 3.5 *Let Assumptions 2.1-2.3 and 3.1 hold. Let the sequence* $\{(\boldsymbol{u}(k), \boldsymbol{v}(k), \boldsymbol{y}(k), \boldsymbol{\mu}(k))\}$ *be generated by Algorithm 3.2 and* $(\boldsymbol{u}^\star, \boldsymbol{v}^\star, \boldsymbol{y}^\star, \boldsymbol{\mu}^\star)$ *be a saddle point of* $L_2(\boldsymbol{u}, \boldsymbol{v}, \boldsymbol{y}, \boldsymbol{\mu})$ *as defined in (3.3). Furthermore, let* $V_2(k)$ *be defined in (3.51). If* $\eta_i^j \in (0, 1)$*, for all* $j \in \mathcal{N}_i$ *and* $i \in \mathcal{N}$*, then* $\{V_1(k)\}$ *is a monotonically non-increasing sequence and the following inequality holds:*

$$\begin{aligned}V_2(k+1) - V_2(k) \leq &-\frac{1}{2}\|\boldsymbol{r}(k+1) + \boldsymbol{y}(k+1) - \boldsymbol{y}(k)\|_2^2 - \sum_{i\in\mathcal{N}} \sigma_i\|\boldsymbol{u}_i(k+1) - \boldsymbol{u}_i^\star\|_2^2 \\ &- \sum_{i\in\mathcal{N}}\sum_{j\in\mathcal{N}_i} \frac{1-\eta_i^j}{2}\|\boldsymbol{r}_i^j(k+1)\|_2^2 \leq 0. \end{aligned} \tag{3.52}$$

Proof First, using the definition of $\boldsymbol{r}(k)$ in (3.39) and the dual variable update in (3.38), it holds that

$$\begin{aligned}\frac{1}{2\eta_i^j}\|\boldsymbol{\mu}_i^j(k+1) - \boldsymbol{\mu}_i^{j\star}\|_2^2 = &\frac{1}{2\eta_i^j}\|\boldsymbol{\mu}_i^j(k) - \boldsymbol{\mu}_i^{j\star}\|_2^2 + \frac{\eta_i^j}{2}\|\boldsymbol{r}_i^j(k+1)\|_2^2 \\ &+ \langle \boldsymbol{\mu}_i^j(k) - \boldsymbol{\mu}_i^{j\star}, \boldsymbol{r}_i^j(k+1)\rangle. \end{aligned} \tag{3.53}$$

Then, the difference between $V_2(k+1)$ and $V_2(k)$, for each $k \in \mathbb{Z}_{\geq 0}$, is stated as follows:

$$\begin{aligned}
V_2(k+1) - V_2(k) &= \frac{1}{2}\|\boldsymbol{\mu}(k+1) - \boldsymbol{\mu}^\star\|_H^2 + \frac{1}{2}\|\boldsymbol{y}(k+1) - \boldsymbol{y}^\star\|_2^2 \\
&\quad - \frac{1}{2}\|\boldsymbol{\mu}(k) - \boldsymbol{\mu}^\star\|_H^2 - \frac{1}{2}\|\boldsymbol{y}(k) - \boldsymbol{y}^\star\|_2^2 \\
&= \frac{1}{2}\|\boldsymbol{r}(k+1)\|_{H^{-1}}^2 + \langle \boldsymbol{\mu}(k) - \boldsymbol{\mu}^\star, \boldsymbol{r}(k+1)\rangle \\
&\quad + \frac{1}{2}\|\boldsymbol{y}(k+1) - \boldsymbol{y}^\star\|_2^2 - \frac{1}{2}\|\boldsymbol{y}(k) - \boldsymbol{y}^\star\|_2^2 \\
&\leq -\frac{1}{2}\|\boldsymbol{r}(k+1) + \boldsymbol{y}(k+1) - \boldsymbol{y}(k)\|_2^2 - \sum_{i \in \mathcal{N}} \sigma_i \|\boldsymbol{u}_i(k+1) - \boldsymbol{u}_i^\star\|_2^2 \\
&\quad - \sum_{i \in \mathcal{N}} \sum_{j \in \mathcal{N}_i} \frac{1 - \eta_i^j}{2} \|\boldsymbol{r}_i^j(k+1)\|_2^2 \leq 0,
\end{aligned} \tag{3.54}$$

where the second equality is obtained by using the relation in (3.53), the first inequality is obtained from the inequality (3.40), and the last inequality holds since $\eta_i^j \in (0, 1)$, for all $j \in \mathcal{N}_i$ and $i \in \mathcal{N}$.

Finally the convergence properties of Algorithm 3.2 are shown in the following theorem.

Theorem 3.2 *Let Assumptions 2.1-2.3 and 3.1 hold. Furthermore, let the sequence* $\{(\boldsymbol{u}(k), \boldsymbol{v}(k), \boldsymbol{y}(k), \boldsymbol{\mu}(k))\}$ *be generated by Algorithm 3.2. If* $\eta_i^j \in (0, 1)$, *for all* $j \in \mathcal{N}$ *and* $i \in \mathcal{N}$, *then,*

1. *(Feasibility)* $\lim_{k\to\infty} \|\boldsymbol{v}_i^j(k) + \boldsymbol{v}_j^i(k)\|_2^2 = 0$, *for all* $j \in \mathcal{N}_i$ *and* $i \in \mathcal{N}$,
2. *(Primal and dual variable convergence) There exists a saddle point of* $L_2(\boldsymbol{u}, \boldsymbol{v}, \boldsymbol{y}, \boldsymbol{\mu})$ *(see (3.29)), denoted by* $(\boldsymbol{u}^\star, \boldsymbol{v}^\star, \boldsymbol{y}^\star, \boldsymbol{\mu}^\star)$, *such that* $\lim_{k\to\infty} \boldsymbol{u}(k) = \boldsymbol{u}^\star$, $\lim_{k\to\infty} \boldsymbol{v}(k) = \boldsymbol{v}^\star$, $\lim_{k\to\infty} \boldsymbol{y}(k) = \boldsymbol{y}^\star$, *and* $\lim_{k\to\infty} \boldsymbol{\mu}(k) = \boldsymbol{\mu}^\star$. □

Proof The proof follows the same steps of the proof of Theorem 3.1. First, rearranging and iterating the first inequality in (3.52) over $\ell = 0, 1, \ldots, k$ implies that

$$\lim_{k\to\infty} \|\boldsymbol{r}(k+1) + \boldsymbol{y}(k+1) - \boldsymbol{y}(k)\|_2^2 = 0, \tag{3.55}$$

$$\lim_{k\to\infty} \sigma_i \|\boldsymbol{u}_i(k+1) - \boldsymbol{u}_i^\star\|_2^2 = 0, \quad \forall i \in \mathcal{N}, \tag{3.56}$$

$$\lim_{k\to\infty} \|\boldsymbol{r}(k+1)\|_2^2 = \lim_{k\to\infty} \|\boldsymbol{v}(k+1) - \boldsymbol{y}(k+1)\|_2^2 = 0. \tag{3.57}$$

Moreover, by (3.55) and (3.57), it holds that

$$\lim_{k\to\infty} \|\boldsymbol{y}(k+1) - \boldsymbol{y}(k)\|_2^2 = 0. \tag{3.58}$$

Additionally, due to the update rule in (3.37), $y_i^j(k) = -y_j^i(k)$, for all $j \in \mathcal{N}_i$, $i \in \mathcal{N}$, and $k \in \mathbb{Z}_{\geq 1}$. Based on this fact and the relation in (3.57), it also holds that

$$\lim_{k\to\infty} \|v_i^j(k+1) + v_j^i(k+1)\|_2^2 = 0, \ \forall j \in \mathcal{N}_i, \ \forall i \in \mathcal{N}. \tag{3.59}$$

Furthermore, based on (3.52), the sequences $\{y(k)\}$ and $\{\mu(k))\}$ are bounded and have accumulation points. Additionally, based on (3.57), $\{v(k)\}$ is also bounded and has accumulation points, which are equal to that of $\{y(k)\}$. Now, consider a convergent subsequence $\{(v(k_\ell), y(k_\ell), \mu(k_\ell))\}$ and its limit point, denoted by (v^a, y^a, μ^a), where $v^a = y^a$. Then, it will be shown that $(u^\star, v^a, y^a, \mu^a)$ is a saddle point of the Lagrangian function $L_2(u, v, y, \mu)$ as defined in (3.29), i.e., $(u^\star, v^a, y^a, \mu^a)$ satisfies the saddle point inequalities in (3.31).

Based on (3.57), it holds that $v_i^{ja} - y_i^{ja} = \lim_{\ell\to\infty}(v_i^j(k_\ell) - y_i^j(k_\ell)) = 0$, for all $j \in \mathcal{N}_i$ and $i \in \mathcal{N}$. Therefore, for any $\mu \in \mathbb{R}^{\sum_{i\in\mathcal{N}} h|\mathcal{N}_i|}$, $L_2(u^\star, v^a, y^a, \mu) = L_2(u^\star, v^a, y^a, \mu^a)$, satisfying the first inequality in (3.31). Now, by substituting k with k_ℓ in (3.36), for all $i \in \mathcal{N}$, and taking the limit as ℓ goes to infinity, it holds that

$$\begin{aligned}
(u^\star, v^a) &= \lim_{\ell\to\infty} \arg \min_{(u_i,v_i)\in\mathcal{L}_i, i\in\mathcal{N}} \sum_{i\in\mathcal{N}} \Big(f_i^{l}(u_i) + f_i^{c}(v_i) \\
&\quad + \sum_{j\in\mathcal{N}_i} \left(\langle \mu_i^j(k_\ell), v_i^j\rangle + \|v_i^j - y_i^j(k_\ell)\|_2^2\right)\Big) \\
&= \arg \min_{(u_i,v_i)\in\mathcal{L}_i, i\in\mathcal{N}} \sum_{i\in\mathcal{N}} \Big(f_i^{l}(u_i) + f_i^{c}(v_i) + \sum_{j\in\mathcal{N}_i}\left(\langle \mu_i^{ja}, v_i^j\rangle + \|v_i^j - y_i^{ja}\|_2^2\right)\Big) \\
&= \arg \min_{(u_i,v_i)\in\mathcal{L}_i, i\in\mathcal{N}} \sum_{i\in\mathcal{N}} \Big(f_i^{l}(u_i) + f_i^{c}(v_i) + \sum_{j\in\mathcal{N}_i} \langle \mu_i^{ja}, v_i^j\rangle - \langle \mu_i^{ja}, y_i^{ja}\rangle\Big),
\end{aligned} \tag{3.60}$$

where the last equality is obtained by introducing the constant $-\sum_{i\in\mathcal{N}}\sum_{j\in\mathcal{N}_i}\langle \mu_i^{ja}, y_i^{ja}\rangle$ and since the quadratic term $\sum_{i\in\mathcal{N}}\sum_{j\in\mathcal{N}_i}\|v_i^j - y_i^{ja}\|_2^2$ is 0 at $(u^\star, v^a)$ due to (3.57). Now, note that (v^a, y^a) minimizes $\sum_{i\in\mathcal{N}}\sum_{j\in\mathcal{N}_i}\|v_i^j - y_i^j\|_2^2$ since $v_i^{ja} - y_i^{ja} = 0$, for all $j \in \mathcal{N}_i$ and $i \in \mathcal{N}$. Therefore, $(u^\star, v^a, y^a)$ minimizes $L_2(u, v, y, \mu^a)$, implying the satisfaction of the second inequality in (3.31). Hence, it can be concluded that $(u^\star, v^a, y^a, \mu^a)$ is a saddle point of $L_2(u, v, y, \mu)$. Finally, set $y^\star = y^a$ and $\mu^\star = \mu^a$ in $V_2(k)$ (see (3.51)). Since the subsequence of $V_2(k_\ell)$ converges to 0 and due to the monotonicity of $V_2(k)$, the entire sequence $\{(u(k), y(k), \mu(k))\}$ converges to (v^a, y^a, μ^a).

3.4 Summary

In this chapter, two distributed algorithms that are based on the augmented Lagrangian approach are proposed to solve Problem (3.1). The comparison between the two algorithms is given as follows:

- *Problem augmentation.* The first step of designing both algorithms is augmenting Problem (3.1). While in the design of Algorithm 3.1, the auxiliary terms are based on the coupling constraints (3.1b), in the design of Algorithm 3.2, auxiliary variables and constraints are introduced, and the augmentation is based on them.
- *Algorithm design.* Each algorithm is derived based on the steps of solving its dual problem, where primal variables are updated by computing the attainers of local optimization problems and dual variables are updated using the gradient ascent step. Regarding the computation of the primal variables, Algorithm 3.1 considers a parametrization of its dual function and uses a convex combination step to update the coupled variables. Differently, Algorithm 3.2 alternatively solves two decomposable optimizations to compute the primal variables.
- *Information exchange.* At each iteration, Algorithm 3.1 in the general form requires two rounds of communication (steps 3 and 5) whereas Algorithm 3.2 only requires one round of communication. However, the amount of information exchanged per iteration is the same. Additionally, as mentioned in Remark 3.1, with a proper initialization, Algorithm 3.1 only needs one round of communication per iteration.
- *Convergence condition.* The sufficient conditions of the step sizes η_i^j, for all $j \in \mathcal{N}_i$ and $i \in \mathcal{N}$, that guarantee the convergence of both algorithms are also different. While Algorithm 3.1 requires an agreement between η_i^j and η_j^i, when the network uses Algorithm 3.2, each subsystem can choose η_i^j independently. Moreover, the range in which the value η_i^j can be chosen is also different between both algorithms.
- *Convergence analysis.* As shown in Theorems 3.1 and 3.2 that both algorithms produce a sequence that converges to a saddle point. The convergence analysis of both algorithms is essentially the same, where the optimality conditions of the optimization problems involved in the algorithms are evaluated and then a suitable Lyapunov function is proposed and used to show the convergence. Note that the Lyapunov functions used in the convergence analysis of both algorithms are different.

Both algorithms are scalable since each local controller only deals with a relatively small dimension decisions, which are associated with the nodes of its subsystem, and it also only communicates with its neighbors. Furthermore, both algorithms require a fixed communication network where each pair of neighboring subsystems must be able to communicate. Moreover, all subsystems must perform the updates synchronously. These conditions are stated in Assumption 3.1. In the next two chapters, the communication network will be considered not as ideal as in Assumption 3.1. In practice, the communication network might be time-varying, e.g., due to link failures. Chapter 4 discusses information exchange protocols that can be used by any distributed method when communication link failures occur whereas Chap. 5

focuses specifically on modifying Algorithms 3.1 and 3.2 such that they still work when the communication network is time-varying and the updates are performed asynchronously by the subsystems.

References

1. Bertsekas DP (2008) Extended monotropic programming and duality. J Optim Theory Appl 139(2):209–225
2. Boyd S, Parikh N, Chu E, Peleato B, Eckstein J (2011) Distributed optimization and statistical learning via the alternating direction method of multipliers. Found Trends Mach Learn 3(1):1–122
3. Boyd S, Vandenberghe L (2010) Convex optimization, vol 25. Cambridge University Press, Cambridge
4. Bertsekas DP (1995) Nonlinear programming. Athena Scientific
5. Larsen GKH, van Foreest ND, Scherpen JMA (2014) Distributed MPC applied to a network of households with micro-CHP and heat storage. IEEE Trans Smart Grid 5(4):2106–2114
6. Kraning M, Chu E, Lavaei J, Boyd S (2014) Dynamic network energy management via proximal message passing. Found Trends in Optim 1(2):73–126
7. Wang T, O'Neill D, Kamath H (2015) Dynamic control and optimization of distributed energy resources in a microgrid. IEEE Trans Smart Grid 6(6):2884–2894
8. Hans CA, Braun P, Raisch, J, Grune L, Reincke-Collon C (2019) Hierarchical distributed model predictive control of interconnected microgrids. IEEE Trans Sustain Energy 10(1):407–416
9. Nedić A, Ozdaglar A (2009) Distributed subgradient methods for multi-agent optimization. IEEE Trans Autom Control 54(1):48–61
10. Nedić A, Ozdaglar A, Parrilo PA (2010) Constrained consensus and optimization in multi-agent networks. IEEE Trans Autom Control 55(4):922–938
11. Chang T (2016) A proximal dual consensus ADMM method for multi-agent constrained optimization. IEEE Trans Signal Process 64(14):3719–3734
12. Chatzipanagiotis N, Dentcheva D, Zavlanos MM (2015) An augmented Lagrangian method for distributed optimization. Math Program 152(1):405–434
13. Chatzipanagiotis N, Zavlanos MM (2016) A distributed algorithm for convex constrained optimization under noise. IEEE Trans Autom Control 61(9):2496–2511
14. Lee S, Chatzipanagiotis N, Zavlanos MM (2018) Complexity certification of a distributed augmented Lagrangian method. IEEE Trans Autom Control 63(3):827–834
15. Mulvey JM, Ruszczyński A (1992) A diagonal quadratic approximation method for large scale linear programs. Oper Res Lett 12(4):205–215
16. Berger A, Mulvey J, Ruszczyński A (1994) An extension of the DQA algorithm to convex stochastic programs. SIAM J Optim 4(4):735–753
17. Ruszczyński A (1995) On convergence of an augmented Lagrangian decomposition method for sparse convex optimization. Math Oper Res 20(3):634–656
18. Bertsekas DP, Tsitsiklis JN (1997) Parallel and distributed computation: numerical methods. Athena Scientific
19. Wei E, Ozdaglar A (2013) On the O(1/k) convergence of asynchronous distributed alternating direction method of multipliers, pp 1–30. arXiv:1307.8254
20. Chang T, Hong M, Liao W, Wang X (2016) Asynchronous distributed ADMM for large-scale optimization—part I: Algorithm and convergence analysis. IEEE Trans Signal Process 64(12):3118–3130
21. Beck A (2017) First-order methods in optimization. Society for Industrial and Applied Mathematics
22. Nedić A (2008) Lecture notes optimization I. Hamilton Institute

23. Uribe CA, Lee S, Gasnikov A, Nedić A (2018) A dual approach for optimal algorithms in distributed optimization over networks, pp 1–37. arXiv:1809.00710
24. Nesterov Y, Nemirovskii A (1994) Interior-point polynomial algorithms in convex programming, vol 13. SIAM
25. Nesterov Y (2013) Introductory lectures on convex optimization: a basic course, vol 87. Springer Science & Business Media, Berlin

Chapter 4
Mitigating Communication Failures in Distributed MPC Schemes

Information sharing among local controllers is a key feature of any distributed model predictive control (DMPC) strategy. This chapter addresses the problem of communication failures in DMPC strategies and proposes a distributed solution to cope with them. The proposal consists in an information-exchange protocol that is based on consensus. By applying this protocol as a complementary plug-in to a DMPC strategy, the controllers improve the resiliency against communication failures and relax the requirements of the communication network. Furthermore, a discussion on the selection criteria of the information-sharing network and a reconfiguration algorithm, which is a contingency procedure to maintain the connectivity of the network, are also presented. In order to demonstrate the performance and advantages of the proposed approach when it is applied to a DMPC strategy, a case study of the economic dispatch problem is provided. Note that the explanation in this chapter is not restricted to the scheme with the distributed algorithms presented in Chap. 3. Instead, the proposed protocol can be applied to a broad class of DMPC strategies.

4.1 Introduction

Different DMPC approaches require different communication structures as well as different ways to exchange information among local controllers [1]. Some DMPC approaches require local controllers to share information iteratively while others require that the information exchange is made only once at each time instant. In terms of their communication networks, some approaches, such as the algorithms presented in Chap. 3 and those discussed in [2–4], require a neighbor-to-neighbor communication while other approaches, e.g., [5, 6], require all controllers to exchange information with all the others. In any way, the information-sharing network, through which the exchange of information occurs, is important for systems that use a DMPC approach.

W. W. Ananduta, *Non-centralized Optimization-Based Control Schemes for Large-Scale Energy Systems*, Springer Theses,
https://doi.org/10.1007/978-3-030-89803-8_4

An information-sharing network may face some problems during the operation of the associated system. Those issues include communication failures (total loss of communication links), delays, and data packet drops [7]. This chapter focuses on the failures (loss of links) of the information-sharing network, implying some local controllers are no longer able to communicate with others. As mentioned in [7] and later discussed, communication failures may lead to severe problems such as the inability of the controllers to compute control inputs or the suboptimality of the solutions.

Some recent literature has addressed the problem of communication failures in a distributed control strategy, in particular DMPC. The authors of [8] analyse the performance degradation of a DMPC strategy that is based on Nash optimality during such failures while assuming that the algorithm is convergent. The authors of [9] propose a scheme where the subsystems assume that their neighbors take null control actions during communication failures. In [10], the authors develop a methodology to extend the DMPC strategy that is proposed in [11] such that it can cope with communication failures. The methodology involves substituting the coupled constraints with tube-based constraints that restrict the control inputs. Furthermore, Kumar et al. [12] propose to add an observer for a robust DMPC strategy. Hence, during a communication failure, the state bounds are estimated by the observer and are posed as extra constraints into the DMPC design. Additionally, a resilient information-sharing network architecture for distributed frequency regulation is proposed in [13]. Moreover, the controller in [13] adopts a *zero-bias* control strategy and allows other subsystems to stabilize themselves.

The aforementioned contributions only improve some DMPC strategies to tackle communication failures in a way that they specifically add or modify the algorithms. Therefore, they limit the application of the solution only to the DMPC strategies that are discussed in those papers. In this regard, different from the works previously discussed, this chapter proposes a communication protocol that can be applied regardless the DMPC strategy that is used to control the system. The protocol is an iterative algorithm that requires local controllers to communicate at each iteration until the information that is received converges to the correct value. It is based on the distributed consensus protocol [14–16] and can also be perceived as distributed projection dynamics (DPD) [17, 18].

The remaining of the chapter is structured as follows. Section 4.2 provides an analysis of the impact of communication failures in a DMPC strategy. Then, Sect. 4.3 presents the consensus-based information-exchange protocol. Afterward, some criteria to select a suitable information-sharing network for the proposed protocol and a network reconfiguration algorithm that supplements the protocol are proposed in Sect. 4.4. Furthermore, the advantages of the proposed protocol in the economic dispatch problem of prosumers are shown in Sect. 4.5. Finally, some concluding remarks are drawn in Sect. 4.6.

4.2 Impact of Communication Failures

This section is dedicated to show, by an example, what could happen when there are failures in the information-sharing network, i.e., at least one communication link fails for a certain time slot. During this period, the affected subsystems cannot communicate with each other while employing a DMPC strategy. For this example, consider an energy system with a network of subsystems $\mathcal{P}$, which apply a DMPC-based economic dispatch. Moreover, consider a set of neighboring subsystems, denoted by $\mathcal{P}^{\mathrm{lf}} = \{p, q \in \mathcal{P} : \{p, q\} \in \mathcal{E}_t^{\mathrm{s}}\} \subset \mathcal{P}$ suffer a communication link failure, i.e., their communication links connecting each other fail at time step t^{lf}, i.e., $\{p, q\} \notin \mathcal{E}^{\mathrm{c}}_{t^{\mathrm{lf}}}$, for all $p, q \in \mathcal{P}^{\mathrm{lf}}$. This implies that at time step t^{lf}, the subsystems in $\mathcal{P}^{\mathrm{lf}}$ cannot exchange information among each other although it is necessary to do so to perform the distributed algorithm, say Algorithm 3.1 or 3.2. In this regard, it is assumed that these subsystems use the old information that is available from the previous time step, $t^{\mathrm{lf}} - 1$, and consider the unknown information to be null, as in [9, 13]. This is precisely stated in Assumption 4.1.

Assumption 4.1 Consider that two neighboring subsystems p and q do not have a communication link at time step t^{lf}, i.e., $\{p, q\} \in \mathcal{E}_t^{\mathrm{s}}$ but $\{p, q\} \notin \mathcal{E}^{\mathrm{c}}_{t^{\mathrm{lf}}}$. Therefore, at t^{lf}, subsystem p considers the decision of subsystem q, i.e., $\boldsymbol{v}^i_{j,t^{\mathrm{lf}}}$ for all $j \in \mathcal{M}_{q,t^{\mathrm{lf}}}$ and $i \in \mathcal{M}_{p,t^{\mathrm{lf}}}$, denoted by $\check{\boldsymbol{v}}^i_{j,t^{\mathrm{lf}}} = \mathrm{col}(\{\check{v}^i_{j,\tau}\}_{\tau\in\mathcal{T}})$, as follows:

$$\check{v}^i_{j,\tau} = \begin{cases} v^i_{j,\tau|t^{\mathrm{lf}}-1}, & \tau = t^{\mathrm{lf}}, \ldots, t^{\mathrm{lf}} + h - 2, \\ 0, & \tau = t^{\mathrm{lf}} + h - 1, \end{cases}$$

and vice versa. Note that the subscript $\tau | t^{\mathrm{lf}} - 1$ indicates that the decision is computed at $t^{\mathrm{lf}} - 1$. □

By considering Assumption 4.1, the subsystems in $\mathcal{P}^{\mathrm{lf}}$ can immediately compute some of their coupled decisions, i.e., $\boldsymbol{v}^j_{i,t^{\mathrm{lf}}} = -\check{\boldsymbol{v}}^i_{j,t^{\mathrm{lf}}}$, for all $j \in \mathcal{M}_{q,t}$ and $i \in \mathcal{M}_{p,t}$, where $p \neq q$ and $p, q \in \mathcal{P}^{\mathrm{lf}}$. This implies that, instead of solving Problem (2.15), the subsystems in the network essentially try solve the following problem:

$$\underset{\{\boldsymbol{u}_{i,t}, \boldsymbol{v}_{i,t}\}_{i\in\mathcal{N}}}{\text{minimize}} \sum_{p\in\mathcal{P}} \sum_{i\in\mathcal{M}_{p,t}} \left(f_i^{\mathrm{l}}(\boldsymbol{u}_{i,t}) + f_i^{\mathrm{c}}(\boldsymbol{v}_{i,t})\right) \tag{4.1a}$$

$$\text{s.t. } (\boldsymbol{u}_{i,t}, \boldsymbol{v}_{i,t}) \in \mathcal{L}_i, \ \forall i \in \mathcal{M}_{p,t}, \ \forall p \in \mathcal{P}, \tag{4.1b}$$

$$\boldsymbol{v}^j_{i,t} + \boldsymbol{v}^i_{j,t} = 0, \quad \forall j \in \mathcal{N}_i \cap \mathcal{M}_{p,t}, \ \forall i \in \mathcal{M}_{p,t}, \ \forall p \in \mathcal{P}, \tag{4.1c}$$

$$\boldsymbol{v}^j_{i,t} + \boldsymbol{v}^i_{j,t} = 0, \quad \forall j \in \mathcal{N}_i \backslash \mathcal{M}_{p,t}, \ \forall i \in \mathcal{M}_{p,t}, \ \forall p \in \mathcal{P} \backslash \mathcal{P}^{\mathrm{lf}}, \tag{4.1d}$$

$$\boldsymbol{v}^j_{i,t} + \boldsymbol{v}^i_{j,t} = 0, \quad \forall j \in \mathcal{N}_i \cap \mathcal{M}_{q,t}, \ \forall q \in \mathcal{P} \backslash \mathcal{P}^{\mathrm{lf}}, \ \forall i \in \mathcal{M}_{p,t}, \ \forall p \in \mathcal{P}^{\mathrm{lf}}, \tag{4.1e}$$

$$\boldsymbol{v}^j_{i,t} + \check{\boldsymbol{v}}^i_{j,t} = 0, \quad \forall j \in \mathcal{N}_i \cap \mathcal{M}_{q,t}, \ \forall q \in \mathcal{P}^{\mathrm{lf}}, \ \forall i \in \mathcal{M}_{p,t}, \ \forall p \in \mathcal{P}^{\mathrm{lf}}, \tag{4.1f}$$

where $t = t^{\mathrm{lf}}$. Note that the equality constraints in (4.1f) are local constraints and differentiate Problem (4.1) from Problem (2.15). The DMPC scheme in [9], which apply Assumption 4.1 during failures, also assumes that null control input is a feasible solution. However, the latter assumption does not always hold in general. Therefore, Problem (4.1) might actually be infeasible. If feasible solutions to Problem (4.1) exist, then they are also feasible solutions to Problem (2.15). However, these solutions might not be an optimal solution to Problem (2.15). Hence, increasing the resiliency of the communication infrastructure is important to avoid the occurrence such problems.

4.3 Information-Exchange Protocol

This section discusses a proposal to improve the resiliency of the information-sharing network. Mainly, it is an application of the consensus algorithm as an information-exchange protocol for the DMPC schemes. First, the proposed protocol is presented and then the advantages of the protocol is explained.

Recall the communication network of subsystems at a fixed time step t as the information-sharing graph $\mathcal{G}^{\mathrm{c}} = (\mathcal{P}, \mathcal{E}^{\mathrm{c}})$, where the subscript t is dropped for notational simplicity. Furthermore, consider that some supra-nodes (subsystems) require information denoted by $\boldsymbol{\theta}_r = \mathrm{col}(\{\theta_r^l\}_{l=1}^{n_r^{\theta}}) \in \mathbb{R}^{n_r^{\theta}}$, from subsystem $r \in \mathcal{P}$. Depending on the DMPC strategy applied to the system and the couplings in the system, $\boldsymbol{\theta}_r$ may consist of either state or input information, which is required by some or all other nodes to run the DMPC algorithm. In the distributed algorithms presented in Chap. 3, the information that must be shared is some coupled variables and dual variables. Therefore, there exists a sub-graph $\mathcal{G}_r^{\mathrm{c}} = (\mathcal{P}_r, \mathcal{E}_r^{\mathrm{c}}) \subseteq \mathcal{G}^{\mathrm{c}}$, where $\mathcal{P}_r \subseteq \mathcal{P}$ is the set that consists of but not limited to subsystem r and the other subsystems that require $\boldsymbol{\theta}_r$, while $\mathcal{E}_r^{\mathrm{c}} \subseteq \mathcal{E}^{\mathrm{c}}$ is the set of communication links available that connect $\mathcal{P}_r$. For instance, in the DAL and ADMM-based distributed algorithms, which require neighbor-to-neighbor communication, some neighbors $q \in \mathcal{N}_r^{\mathrm{s}}$ might be included in $\mathcal{P}_r$ while the other neighbors are included in the other information-sharing subgraphs. In order to apply the protocol, the following assumptions must hold.

Assumption 4.2 The undirected sub-graph $\mathcal{G}_r^{\mathrm{c}}$ is connected. □

Assumption 4.3 Subsystem r, which sends the information, has prior knowledge of $|\mathcal{P}_r|$, while all $p \in \mathcal{P}_r \backslash \{r\}$ know n_r^{θ}. □

Remark 4.1 Following Assumption 4.2, it is possible that there are some nodes in $\mathcal{P}_r$ that do not need the information $\boldsymbol{\theta}_r$, but they are required as intermediate nodes in order to ensure the connectivity of $\mathcal{G}_r^{\mathrm{c}}$. □

The proposed information-exchange protocol is based on the following consensus dynamics:

$$\dot{\boldsymbol{\psi}}_p(\kappa) = \sum_{q \in \mathcal{N}_p^{\mathrm{c}}} \alpha(\boldsymbol{\psi}_q(\kappa) - \boldsymbol{\psi}_p(\kappa)), \quad \forall p \in \mathcal{P}_r, \tag{4.2}$$

where $\boldsymbol{\psi}_p \in \mathbb{R}^{n_r^\theta}$ is the information state of subsystem $p \in \mathcal{P}_r$. Moreover, the index κ is used to denote continuous time, $\mathcal{N}_p^c$ is the set of subsystems that are the neighbors of subsystem p in $\mathcal{G}_r^c$, i.e., $\mathcal{N}_p^c = \{q \in \mathcal{P}_r : \{p, q\} \in \mathcal{E}_r^c\}$, and $\alpha \in \mathbb{R}_{>0}$ is a constant gain. Furthermore, the information states of all nodes in $\mathcal{P}_r$ are initialized by

$$\boldsymbol{\psi}_p(0) = \begin{cases} |\mathcal{P}_r|\boldsymbol{\theta}_r, & p = r, \\ \mathbf{0}_{n_r^\theta}, & \text{otherwise.} \end{cases} \tag{4.3}$$

Note that the vectors $\boldsymbol{\psi}_p$ for all $p \in \mathcal{P}_r$ are dedicated only for the transmission of information $\boldsymbol{\theta}_r$, which comes from subsystem r. Thus, each local controller needs to allocate a different data storage for acquiring another information, i.e., if there are more than one source nodes.

Proposition 4.1 *Suppose that Assumptions 4.2 and 4.3 hold. Under the dynamics (4.2) and by initializing the information states as in (4.3),* $\lim_{\kappa \to \infty} \boldsymbol{\psi}_p(\kappa) = \boldsymbol{\theta}_r$, *for each* $p \in \mathcal{P}_r$.

Proof Let $\boldsymbol{\psi}(\kappa) = \mathrm{col}(\{\boldsymbol{\psi}_p(\kappa)\}_{p \in \mathcal{P}_r}) \in \mathbb{R}^{|\mathcal{P}_r|n_r^\theta}$ be the information state of the overall system. There exists a permutation matrix $\Phi \in \mathbb{R}^{|\mathcal{P}_r|n_r^\theta \times |\mathcal{P}_r|n_r^\theta}$ such that $\check{\boldsymbol{\psi}}(\kappa) = \Phi\boldsymbol{\psi}(\kappa)$, where $\check{\boldsymbol{\psi}}(\kappa) = \mathrm{col}(\{\boldsymbol{\psi}^l\}_{l=1}^{n_r^\theta})$ and $\boldsymbol{\psi}^l(\kappa) = \mathrm{col}(\{\psi_p^l\}_{p \in \mathcal{P}_r}) \in \mathbb{R}^{|\mathcal{P}_r|}$, for all $l = \{1, \ldots, n_r^\theta\}$. Based on (4.2), the dynamics of $\check{\boldsymbol{\psi}}(\kappa)$ are

$$\dot{\check{\boldsymbol{\psi}}}(\kappa) = -\left(I_{n_r^\theta} \otimes L^{\mathrm{g}}(\mathcal{G}_r^c)\right) \check{\boldsymbol{\psi}}(\kappa),$$

where $L^{\mathrm{g}}(\mathcal{G}_r^c)$ denotes the graph Laplacian of $\mathcal{G}_r^c$. The eigenvalues of $I_{n_r^\theta} \otimes L^{\mathrm{g}}(\mathcal{G}_r^c)$ are the same as the eigenvalues of $L^{\mathrm{g}}(\mathcal{G}_r^c)$ with the algebraic multiplicity of n_r^θ. Hence, the claimed statement follows from the convergence analysis of the consensus dynamics as explained in [14, pp. 46].

Remark 4.2 The convergence rate of the protocol depends on the structure of the network and is indicated by the second smallest eigenvalue of the Laplacian of the graph [15]. □

Remark 4.3 Although the information states asymptotically converge to $\boldsymbol{\theta}_r$, for all $p \in \mathcal{P}_r$, in practice, sufficiently similar information can be recovered in a finite time. □

Two main advantages of the proposed information-exchange protocol are highlighted as follows. Firstly, it relaxes some assumptions that are required by most of DMPC strategies. Mainly, it relaxes the requirement of the information-sharing network topology. As discussed in Sect. 4.1, different DMPC strategies might have different requirements regarding the information-sharing network topology supposing that the information is directly exchanged as required. However, by employing the proposed protocol, these requirements are relaxed such that $\mathcal{G}_t^c$ does not have to fulfill certain topological structure. Instead, only the connectivity of $\mathcal{G}_t^c$ is necessary

(Assumption 4.2). Secondly, this protocol also enhances the resiliency of DMPC-type controllers against communication failures. According to Assumption 4.2, the information can still be exchanged although some links of the network $\mathcal{G}_t^c$ fail as long as the network is still connected. Therefore, to some extent of link failures, a DMPC strategy that uses the proposed protocol to exchange information can still be performed.

The advantages provided by the protocol also come with some costs, which are extra computation and communication, in terms of the amount of data that is exchanged. This is due to the fact that all subsystems should reach consensus by iteratively exchanging information and applying (4.2) before obtaining the correct information from their neighbors. Therefore, one must ensure that the total time to exchange information using the consensus-based protocol and to compute the control inputs is smaller than the sampling time of the controlled system. In practice, the satisfaction of this assumption depends on the system complexity, i.e., the instrumentation and the other hardware as well as the software, e.g., the optimization solver.

4.4 Graph Selection and Reconfiguration

In this section, first, a discussion on how to choose the information-sharing sub-graph, $\mathcal{G}_r^c$, from the available information-sharing network is provided. Two criteria of selection are the resiliency of the network against communication failures and the convergence rate of the proposed method. It is a direct implication that, when the information-sharing graph has more links (edges) connecting the subsystems, the chance that the graph is still connected when a failure occurs is higher. Furthermore, as stated in Remark 4.2, the second smallest eigenvalue of $L^g(\mathcal{G}_r^c)$ indicates the convergence rate of the protocol in the sense that a larger eigenvalue implies a faster convergence rate. As noted in [15], the second smallest eigenvalue of a sparse Laplacian is relatively small compared to a dense Laplacian. Hence, these criteria lead to the fact that the network should have as many links as possible.

Redundancy is required when dealing with communication failures. Thus, when selecting an information-sharing graph, it is important to consider having redundant links in the information-sharing graph. For instance, a path is not a suitable structure since once a link is disconnected, the graph is disconnected. In that sense, a cyclic graph is more redundant because the proposed method could still be applied when one link of this graph is broken. Furthermore, it is obvious that a complete graph is the most suitable one. On the other hand, the topology of a large-scale system usually has a sparse Laplacian matrix due to the fact that a subsystem is usually only coupled with other closest subsystems. This may imply its information-sharing network has sparse Laplacian as well. However, this network can be decomposed into some sub-graphs that do not have sparse Laplacian, which implies they may have faster convergence rate. By also considering the redundancy criterion, then one may be able to use a smaller yet redundant information-sharing graph.

As an example, consider the information-sharing graph depicted in Fig. 4.1. Its Laplacian has the second smallest eigenvalue of 0.23. However, now consider its sub-graphs that are depicted in Fig. 4.2. Notice that the sub-graphs that are formed

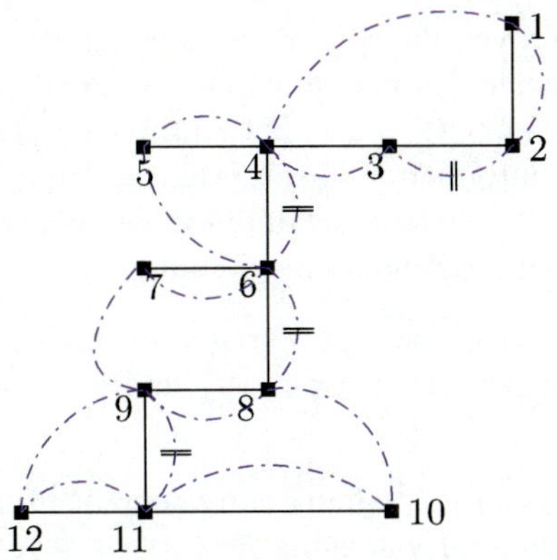

Fig. 4.1 (After [19]) A network of 12 subsystems (squares). The physical network $\mathcal{G}$ and the information-sharing network $\mathcal{G}^c$ are represented by the solid and dash-dotted lines, respectively

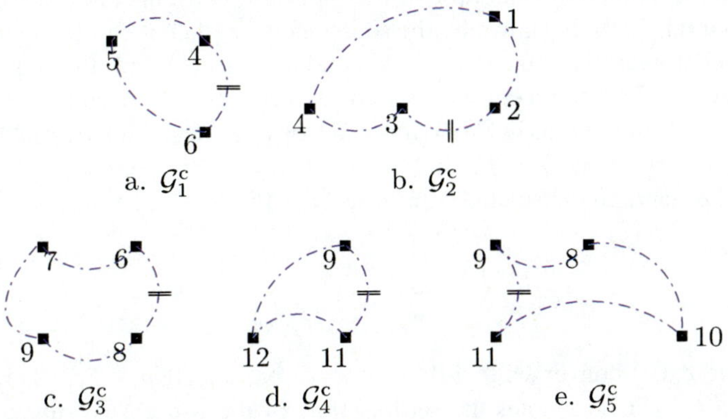

Fig. 4.2 (After [19]) Decomposition of $\mathcal{G}^c$ into 5 smaller connected information sharing sub-graphs

are cyclic, in order to satisfy the redundancy requirement. Among these sub-graphs, the smallest value of the Laplacian second smallest eigenvalues is 2.00. This means that the convergence rate of the proposed method is much faster by using the smaller sub-graphs as the information-sharing graphs among the subsystems. However, the redundancy of the sub-graphs is not as good as the overall graph, $\mathcal{G}^c$. For instance, consider sub-graph $\mathcal{G}_3^c$ (Fig. 4.2c) and suppose that the links $\{7, 9\}$ and $\{6, 7\}$ are broken, then this sub-graph is not connected anymore, which implies the nodes in this sub-graph (6, 7, 8, and 9) cannot exchange information among each other. Nevertheless, the graph $\mathcal{G}^c$ is in fact still connected, allowing the protocol to be applied and all nodes to exchange information when $\mathcal{G}^c$ is used as the information-sharing graph of all information.

Secondly, a distributed algorithm to reconfigure the information-sharing sub-graphs is also proposed. The reconfiguration is a contingency procedure when the information-sharing graphs are disconnected due to the failures, i.e., Assumption

4.2 does not hold. Moreover, the reconfiguration can also be applied even though the graphs are still connected in order to maintain redundancy. Therefore, recall the information-sharing network $\mathcal{G}^{\mathrm{c}} = (\mathcal{P}, \mathcal{E}^{\mathrm{c}})$ and its sub-graph $\mathcal{G}_r^{\mathrm{c}} = (\mathcal{P}_r, \mathcal{E}_r^{\mathrm{c}}) \subseteq \mathcal{G}^{\mathrm{c}}$ that is used to share the information of subsystem r, denoted by $\boldsymbol{\theta}_r$, among the subsystems in $\mathcal{P}_r \subseteq \mathcal{P}$. Now, denote the set of links that fail at time step t^{lf} by $\mathcal{E}^{\mathrm{lf}} \subset \mathcal{E}_r^{\mathrm{c}}$. Moreover, let the following Assumption 4.4 holds.

Assumption 4.4 Each subsystem $p \in \mathcal{P}$ has prior knowledge of its neighbors in the information-sharing network $\mathcal{G}^{\mathrm{c}}$, i.e., $\mathcal{N}_p^{\mathrm{c}}$ and its subset $\tilde{\mathcal{N}}_p^{\mathrm{c}} := \{q : \{p, q\} \in \mathcal{E}^{\mathrm{c}}, \ \forall q \notin \mathcal{P}_r\}$. □

Then, for each link $\{p, q\} \in \mathcal{E}^{\mathrm{lf}}$, find a subsystem that is a neighbor of p and q in $\mathcal{G}^{\mathrm{c}}$ and does not belong to $\mathcal{P}_r$. By defining $\Xi := \{\xi : \xi \in \tilde{\mathcal{N}}_p^{\mathrm{c}} \cap \tilde{\mathcal{M}}_q^{\mathrm{c}}, \ \forall \{p, q\} \in \tilde{\mathcal{E}}^{\mathrm{lf}}\}$ and $\mathcal{E}_\xi^{\mathrm{c}} := \{\{\theta, p\}, \{\theta, q\}, \ \forall \xi \in \Xi, \ \forall \{p, q\} \in \mathcal{E}^{\mathrm{lf}}\}$, the updates of the sub-graph $\mathcal{G}_r^{\mathrm{c}}$ at t^{lf} are $\mathcal{P}_r \leftarrow \mathcal{P}_r \cap \Xi$ and $\mathcal{E}_r^{\mathrm{c}} \leftarrow (\mathcal{E}_r^{\mathrm{c}} \backslash \mathcal{E}^{\mathrm{lf}}) \cap \mathcal{E}_\xi$.

Since the sub-graph $\mathcal{G}_r^{\mathrm{c}}$ is modified, Assumption 4.3 no longer holds. If the number of links that fail, $|\mathcal{E}^{\mathrm{lf}}|$, is known by the source node, r, then it can easily update $|\mathcal{P}_r|$ after the reconfiguration since one node is added for each link that fails. If this is not the case, the distributed consensus algorithm can again be used to recalculate the total number of subsystems in $\tilde{\mathcal{P}}_r$. To this end, consider an auxiliary variable, denoted by $\zeta_p(\kappa)$, for each node $p \in \mathcal{P}_r$. Initialize $\zeta_p(0) = 0$ for $p \neq r$ and $\zeta_r(0) = 1$. By applying the standard distributed consensus [15, 16]

$$\dot{\zeta}_p(\kappa) = \sum_{q \in \mathcal{N}_p^{\mathrm{c}}} \alpha(\xi_j(\kappa) - \xi_i(\kappa)), \tag{4.4}$$

the variable $\zeta_p(\kappa)$ converges to $1/|\mathcal{P}_r|$ as $\kappa \to \infty$, for all $p \in \mathcal{P}_r$. Therefore, in a finite time, κ_s, which denotes the settling time of the consensus, subsystem r can obtain $|\mathcal{P}_r| = \mathrm{rnd}(1/\zeta_r(\kappa_s))$. Note that the whole reconfiguration procedure can be performed in a distributed fashion, i.e., the algorithm only requires each node to have local information of its neighbors (Assumption 4.4) and neighbor-to-neighbor communication. The distributed reconfiguration procedure is stated in Algorithm 4.1.

Algorithm 4.1 Reconfiguration procedure

1. Each subsystem $i \in \mathcal{P}_r$ detects link failures.
2. For all subsystems p and q, where $\{p, q\} \in \tilde{\mathcal{E}}_f$ and $p, q \in \mathcal{P}_r$, detect the links that are failed.
3. For all subsystems p and q, where $\{p, q\} \in \tilde{\mathcal{E}}_f$ and $p, q \in \mathcal{P}_r$, send information of the failed links to the neighbors that do not belong to $\mathcal{P}_r$, e.g., subsystem p sends to $\{o : o \in \tilde{\mathcal{N}}_p^{\mathrm{c}}\}$ and subsystem q sends to $\{o : o \in \tilde{\mathcal{N}}_q^{\mathrm{c}}\}$.
4. For all $\theta \in \Theta$, receive the same information of one failed link from two different neighbors.
5. For all $\theta \in \Theta$, confirm to join the sub-graph $\mathcal{G}_r^{\mathrm{c}}$.
6. Compute the updated $|\mathcal{P}_r|$ using the distributed consensus algorithm.

Remark 4.4 The intersection of $\tilde{\mathcal{N}}_p^{\rm c}$ and $\tilde{\mathcal{N}}_q^{\rm c}$, for any $\{p, q\} \in \tilde{\mathcal{E}}^{\rm lf}$ might be empty. In this case, the associated information-sharing graph cannot be reconfigured. □

4.5 Case Study

In this section, the effectiveness of the proposed protocol is shown through a simulation study under the framework of the economic dispatch problem of interconnected prosumers. Each prosumer is considered as a subsystem. Furthermore, each prosumer has a dispatchable generator, a storage unit, and aggregated load. It is assumed that the cost function is a strictly convex quadratic function as defined in (2.16) and (2.18), and the economic dispatch problem as stated in (2.15) is formulated using the constraints defined by (2.6)–(2.14). The distributed algorithm used to solve their economic dispatch problem (2.15) is the standard dual-ascent [20], which requires neighbor-to-neighbor communication at each iteration. It is assumed that the default protocol of exchanging information is that a subsystem sends and receives information to and from the physical neighbors through the direct communication links available between them. Furthermore, the proposed information-exchange protocol may also be applied to this algorithm to replace the default one. When the proposed protocol is applied, it is assumed that the information shared at the end of the information-exchange steps is similar enough such that it does not affect the convergence of the algorithm.

Numerical simulations are carried out in MATLAB on a PC with 16 GB of RAM and 2.6 GHz Intel core i7. The network consists of 12 prosumers, i.e., $\mathcal{P} = \{1, \ldots, 12\}$. Figure 4.1 depicts the topology of the network, $\mathcal{G}^{\rm s}$, which is time-invariant and the default information-sharing network of the controllers, $\mathcal{G}^{\rm c}$. The simulation time of all simulations is one day with the sampling time of 15 min. The prediction horizon of the DMPC controller is $h = 6$ and the parameters corresponding to each prosumer are shown in Table 4.1. Furthermore, it is assumed that each local controller knows the local load and its forecast over the prediction horizon at each time step.

In order to compare the performance of the proposed method, four scenarios, which are described in Table 4.2, are simulated. The communication failures are defined as follows. At $t \in \{2, 3, 10-13, 40-42, 80-82\}$, the information-sharing network during the scenarios with failure is $\tilde{\mathcal{G}}^{\rm c} = (\mathcal{P}, \tilde{\mathcal{E}}^{\rm c})$, where $\tilde{\mathcal{E}}^{\rm c} = \mathcal{E}^{\rm c} \backslash \{\{2, 3\}, \{4, 6\}, \{9, 11\}\}$ (the failed links are indicated by || in Fig. 4.1). In addition, in Scenarios 2 and 4, the proposed information-exchange protocol uses sub-graphs as shown in Fig. 4.2 as the information-sharing graphs in order to increase the convergence rates. During failures, all information-sharing sub-graphs only lose one link, hence they are still connected.

Scenario 1 is considered as the baseline performance since the controllers produce the global optimal solution. The simulation result of Scenario 2, as seen in Table 4.2, shows that the proposed information-exchange protocol is able to achieve the optimal performance. In Scenario 3, the prosumers that are disconnected cannot exchange

Table 4.1 Parameters of the agents

Parameters	Value	Unit	Sub-system (p)
$u_p^{\mathrm{ch}}, u_p^{\mathrm{dh}}$	50, 50	kW	All
$x_p^{\min}, x_p^{\max}$	20%, 80%	–	All
$u_p^{\mathrm{dg,min}}, u_p^{\mathrm{dg,max}}$	0, 1000	kW	All
$v_p^{q,\max}$, for all $q \in \mathcal{N}_p^{\mathrm{s}}$	100	kW	All
$u_p^{\mathrm{tp,max}}$	2000	kW	All
a_p	0.90	–	All
$c_p^{\mathrm{st}}, c_p^{\mathrm{tp}}, c_p^{\mathrm{t}}$	0.1, 250, 0.1	–	All
c_p^{dg}	2	–	2, 6, 11
	10	–	1, 3–5, 7–10, 12

Table 4.2 The average stage cost during failures

Scenario	Protocol (communication)	Cost (proportional) (%)
1	Default (no failures)	100
2	Consensus-based (no failures)	100
3	Default (with failures)	116
4	Consensus-based (with failures)	100

information using the default protocol. Therefore, these prosumers adopt Assumption 4.1, which means that the unknown information of the neighbors is considered to be null. In this regard, it is considered that feasible solution of the dispatch problem exists under Assumption 4.1. Table 4.2 shows that there is a performance degradation (16% of higher cost than the optimal case) when the prosumers are in Scenario 3. On the other hand, all prosumers are able to obtain the required information with the proposed methodology despite the occurrence of the failures, as expected. Therefore, the optimal solutions can be obtained by the system, as can be seen by comparing the cost of Scenario 4 and that of Scenario 1, which are equal.

4.6 Summary

A methodology to cope with the problem of communication failures in distributed MPC strategies has been proposed. It involves a consensus-based information-exchange protocol and a graph reconfiguration algorithm. A numerical study, which shows the application of this protocol to a DMPC-based economic dispatch scheme of networked prosumers, demonstrates the advantages of the proposed protocol. Next, in Chap. 5, the distributed algorithms presented in Chap. 3 are discussed further and improved such that not only they work under time-varying communication network without the need of using the consensus-based information exchange protocol but also asynchronously. Nevertheless, the main take away from this chapter is the generality of the methodology as it can be applied to other DMPC strategies as well.

References

1. Scattolini R (2009) Architectures for distributed and hierarchical model predictive control—a review. J Process Control 19(5):723–731
2. Farina M, Scattolini R (2012) Distributed predictive control: a non-cooperative algorithm with neighbor-to-neighbor communication for linear systems. Automatica 48(6):1088–1096
3. Conte C, Jones CN, Morari M, Zeilinger MN (2016) Distributed synthesis and stability of cooperative distributed model predictive control for linear systems. Automatica 69:117–125
4. Liu X, Zhang Y, Lee KY (2017) Coordinated distributed MPC for load frequency control of power system with wind farms. IEEE Trans Ind Electron 64(6):5140–5150
5. Stewart BT, Venkat AN, Rawlings JB, Wright SJ, Pannocchia G (2010) Cooperative distributed model predictive control. Syst Control Lett 59(8):460–469
6. Tarisciotti L, Lo Calzo G, Gaeta A, Zanchetta P, Valencia F, Sáez D (2016) A distributed model predictive control strategy for back-to-back converters. IEEE Trans Ind Electron 63(9):5867–5878
7. Ge X, Yang F, Han Q-L (2017) Distributed networked control systems: a brief overview. Inf Sci 380:117–131
8. Li S, Zhang Y, Zhu Q (2005) Nash-optimization enhanced distributed model predictive control applied to the shell benchmark problem. Inf Sci 170(2):329–349
9. Heidarinejad M, Liu J, Muñoz de la Peña D, Davis JF, Christofides PD (2011) Handling communication disruptions in distributed model predictive control. J Process Control 21(1):173–181
10. Schaich RM, Müller MA, Allgöwer F (2014) A distributed model predictive control scheme for networks with communication failure. IFAC Proc Vol 47(3):12004–12009
11. Müller MA, Reble M, Allgöwer F (2012) Cooperative control of dynamically decoupled systems via distributed model predictive control. Int J Robust Nonlinear Control 22(12):1376–1397
12. Kumar D, Al-Gherwi W, Budman H (2016) Robust-distributed MPC tolerant to communication loss. Comput Chem Eng 88:30–38
13. Nazari MH, Grijalva S, Egerstedt M (2016) Distributed frequency regulation in the presence of communication failure. In: Proceedings of IEEE power energy society innovative smart grid technologies conference (ISGT), Minneapolis, USA, pp 1–5
14. Mesbahi M, Egerstedt M (2010) Graph theoretic methods in multiagent networks. Princeton University Press
15. Olfati-Saber R, Murray RM (2004) Consensus problems in networks of agents with switching topology and time-delays. IEEE Trans Autom Control 49(9):1520–1533
16. Olfati-Saber R, Fax JA, Murray RM (2007) Consensus and cooperation in networked multi-agent systems. Proc IEEE 95(1):215–233
17. Barreiro-Gomez J, Obando G, Quijano N (2017) Distributed population dynamics: optimization and control applications. IEEE Trans Syst Man Cybern: Syst 47(2):304–314
18. Quijano N, Ocampo-Martinez C, Barreiro-Gomez J, Obando G, Pantoja A, Mojica-Nava E (2017) The role of population games and evolutionary dynamics in distributed control systems: the advantages of evolutionary game theory. IEEE Control Syst 37(1):70–97

19. Ananduta W, Barreiro-Gomez J, Ocampo-Martinez C, Quijano N (2018) Mitigation of communication failures in distributed model predictive control strategies. IET Control Theory Appl 12(18):2507–2515
20. Boyd S, Parikh N, Chu E, Peleato B, Eckstein J (2011) Distributed optimization and statistical learning via the alternating direction method of multipliers. Found Trends Mach Learn 3(1):1–122

Chapter 5
Distributed Methods with Stochastic Communication

This chapter revisits the distributed algorithms introduced in Chap. 3 and discusses how these algorithms perform over time-varying communication network asynchronously. Therefore, after a brief introduction in Sect. 5.1, the stochastic model of time-varying communication networks and the asynchronicity of performing the updates is explained in Sect. 5.2. Then, by taking into account the stochastic communication model, the modification of the distributed algorithms is presented in Sect. 5.3. Afterwards, the convergence and the convergence rate analysis of the modified algorithms are provided in Sect. 5.4. Furthermore, Sect. 5.5 is devoted to numerical simulations under the framework of energy systems. Finally, Sect. 5.6 concludes this chapter.

5.1 Introduction

As remarked at the end of Chap. 3, in the distributed methods shown in Algorithms 3.1 and 3.2, the agents that are coupled through link-based constraints must always communicate certain information with their neighbors at each iteration. In this chapter, it is considered that the information exchange process might be imperfect, i.e., Assumption 3.1 no longer holds. For instance, some communication link fails or some agents have not finished computing local decisions, thus they cannot send their information at some iterations. In this regard, the communication network is assumed to be randomly time-varying and the updates are asynchronous. Consequently, some adjustments to these distributed algorithms are necessary.

In addition, these issues are also relevant to the application in the energy systems. Since they are critical infrastructures, it is expected the control and management of such systems to be resilient against communication failures. Furthermore, having the ability to update the decisions asynchronously also improves the robustness of

W. W. Ananduta, *Non-centralized Optimization-Based Control Schemes for Large-Scale Energy Systems*, Springer Theses,
https://doi.org/10.1007/978-3-030-89803-8_5

the systems, particularly, when there are delays in the communication. However, distributed methods based on dual decomposition, ADMM, or ADAL for energy management that are proposed in the literature, e.g., [1–5], typically consider a perfect information exchange process, i.e., the necessary information required to execute the updates is available at each iteration. In fact, this is a common and usually implicit assumption considered, as it can also be found in other distributed methods that are proposed in the literature, such as [6–9].

Exhaustive study on consensus-based distributed methods show that this approach works over time-varying communication network [10–13]. In these algorithms, the main assumption considered is that the graph representation of the communication network is jointly (strongly) connected. Furthermore, dual decomposition or ADMM-based algorithms that also incorporate a consensus-based method for different problems over a time-varying communication network have also reported, e.g., in [14–17]. In this thesis, instead of having a jointly connected assumption, the communication network as well as asynchronous updates are modeled as random processes. Therefore, the particular proposal of this chapter is closely related to the work reported in [18, 19]. In [18], an ADMM algorithm that works asynchronously is proposed and the random model of the network is introduced. Differently, in [19], a distributed optimization problem with coupled equality constraints is considered and the ADMM is applied as a method to update the dual variables. It is worth mentioning that the algorithm proposed in [19] also requires connectivity assumption on the communication graph since one objective of the algorithm is to achieve a consensus on the dual variables. A stochastic gradient proximal method for consensus that also considers randomly time-varying network has also been proposed in [20]. Related to energy management problems, the work in [14] proposes a distributed demand response scheme based on the randomized distributed ADMM proposed in [19]. Similarly, random communication failures and asynchronous updates are also considered in [21–23], which discuss distributed ADMM approaches for solving a distributed consensus optimization problem.

5.2 Random Communication

Recall the network of agents $\mathcal{P}$, represented by the undirected graph $\mathcal{G}^{\mathrm{s}} = (\mathcal{P}, \mathcal{E}^{\mathrm{s}})$. The communication network of these agents is represented by the undirected graph $\mathcal{G}^{\mathrm{c}}(k) = (\mathcal{P}, \mathcal{E}^{\mathrm{c}}(k))$, where $\mathcal{E}^{\mathrm{c}}(k) \subseteq \mathcal{E}^{\mathrm{s}}$ denotes the set of communication links that are active at iteration $k-1$, i.e., $\{p, q\} \in \mathcal{E}^{\mathrm{c}}(k)$ means that agents p and q can exchange information between each other at $k-1$. In Assumption 3.1, it is considered that Algorithms 3.1 and 3.2 works under $\mathcal{G}^{\mathrm{c}}(k)$ where $\mathcal{E}^{\mathrm{c}}(k) = \mathcal{E}^{\mathrm{s}}$, for all $k \in \mathbb{Z}_{\geq 0}$. However, this assumption is relaxed here. Particularly, the communication network is modeled as a random graph [18], as follows.

Assumption 5.1 (*Random network*) The set $\mathcal{E}^{\mathrm{c}}(k)$ is a random variable that is independent and identically distributed across iterations. Furthermore, any communication link between two coupled agents p and q, where $\{p, q\} \in \mathcal{E}^{\mathrm{s}}$, is active with a positive probability denoted by β_{pq}, i.e., $\mathbb{P}(\{p, q\} \in \mathcal{E}^{\mathrm{c}}(k)) = \beta_{pq} > 0$. □

Since $\mathcal{E}^{\mathrm{c}}(k)$ is a subset of $\mathcal{E}^{\mathrm{s}}$, it can be considered that there exists a pre-specified communication network, i.e., $\mathcal{E}^{\mathrm{s}}$, and based on Assumption 5.1, the links of this graph are active with a positive probability. For practical reasons, such as the cost of communication infrastructure, having a communication network represented by $\mathcal{E}^{\mathrm{s}}$ is sufficient, since each agent only needs the information from its neighbors to perform the updates based on Algorithms 3.1 and 3.2, i.e., agent p only needs information from its neighbors in $\mathcal{N}_p^{\mathrm{s}}$. It is worth mentioning that the model of the random network used can be considered as a stochastic block model [24], where the probability of each communication link between two neighboring agents (i.e., p and q where $q \in \mathcal{N}_p^{\mathrm{s}}$ and vice versa) being active is positive and can be different between each other, whereas the probability of each edge between two non-neighboring agents (i.e., p and q where $q \notin \mathcal{N}_p^{\mathrm{s}}$ and vice versa) is zero.

Moreover, as can be seen in Algorithms 3.1 and 3.2, each agent must perform the updates and communicate at each iteration, implying the synchronicity that all agents must have. However, this might not be the case in practice. In particular, it might happen that not all agents update their decisions at each iteration. For instance, due to the inability of some agents to solve their local optimization at the end of the iteration. Therefore, the asynchronous updates are also modeled as a random process, as follows. Denote the set of agents that are active and update their primal and dual variables at iteration $k-1$ by $\mathcal{A}(k)$. Then, consider the following assumption.

Assumption 5.2 (*Asynchronous update*) The set $\mathcal{A}(k) \subseteq \mathcal{P}$ is a random variable that is independent and identically distributed across iterations. Moreover, an agent $p \in \mathcal{P}$ is active and updates its primal and dual variables at iteration k with a positive probability denoted by γ_p, i.e., $\mathbb{P}(p \in \mathcal{A}(k)) = \gamma_p > 0$. □

5.3 Stochastic Distributed Algorithms

In this section, Algorithms 3.1 and 3.2 are modified by taking into account the random communication model introduced in Sect. 5.2. Firstly, consider the time-varying nature of the communication network (Assumption 5.1), and allow each agent to perform the updates asynchronously as defined in Assumption 5.2. Based on Assumptions 5.1 and 5.2, an active agent $p \in \mathcal{A}(k)$ can only exchange information to its neighbor $q \in \mathcal{N}_p^{\mathrm{s}}$ if both agents are active and the communication link $\{p, q\}$ is also active. In this regard, for each agent $p \in \mathcal{P}$, denote the set of coupled neighbors with which agent p can exchange information by $\mathcal{A}_p(k) = \{p \in \mathcal{N}_p^{\mathrm{s}} \cap \mathcal{A}(k) : \{p, q\} \in \mathcal{E}^{\mathrm{c}}(k)\}$.

Consider Algorithm 3.1. Under Assumptions 5.1 and 5.2, an active agent $p \in \mathcal{A}(k+1)$ might not have $\boldsymbol{v}_j^i(k)$ and $\boldsymbol{\lambda}_j^i(k)$, for $j \in \mathcal{N}_i \backslash \mathcal{M}_p$ and $i \in \mathcal{M}_p$, to perform steps 1 and 4 at iteration k. Therefore, it needs to track $\boldsymbol{v}_j^i(k)$ and $\boldsymbol{\lambda}_j^i(k)$. In this regard, this information is captured by the auxiliary variables $z_i^j(k)$ and $\boldsymbol{\xi}_i^j(k)$, for all $j \in \mathcal{N}_i$ and $i \in \mathcal{M}_p$, respectively. Based on the preceding considerations, the distributed augmented Lagrangian method is modified as stated in Algorithm 5.1. Observe that the update of $\boldsymbol{v}_i^j(k)$ is also modified even though only local information is required. Now, consider Algorithm 3.2 as well as Assumptions 5.1 and 5.2. Since the information of the neighbors is only used to update the auxiliary variable $\boldsymbol{y}_i(k)$, then the only step that needs a modification is step 3. With a similar adjustment as in the other approach, the ADMM-based method under the random communication model is presented in Algorithm 5.2. As can be seen in Algorithms 5.1 and 5.2, some variables are not always updated at each iteration. Specifically, $\boldsymbol{v}_i^j(k)$ and $\boldsymbol{\lambda}_i^j(k)$ in Algorithm 5.1 as well as $\boldsymbol{y}_i^j(k)$ and $\boldsymbol{\mu}_i^j(k)$ in Algorithm 5.2. The update of these variables depends on the communication availability between agents $\phi(i,\cdot)$ and $\phi(j,\cdot)$, particularly when $\phi(i,\cdot) \neq \phi(j,\cdot)$. This way of updating the variables is intuitively natural and later, in Sect. 5.4, it will be shown that it is the key to ensure the convergence of the produced sequence to an optimal solution.

Remark 5.1 In Algorithm 5.1, to initialize the auxiliary variables $z_i(0)$ and $\boldsymbol{\xi}_i(0)$, for all $i \in \mathcal{N}$, either agent $\phi(i,\cdot) \in \mathcal{P}$ receives $\boldsymbol{v}_j^i(0)$ and $\boldsymbol{\lambda}_j^i(0)$, for all $j \in \mathcal{N}_i \backslash \mathcal{M}_p$ and $i \in \mathcal{M}_p$, from all neighbors $\phi(j,\cdot) \in \mathcal{N}^{\mathrm{s}}_{\phi(i,\cdot)}$ or it is set such that, for each $i \in \mathcal{N}$, $\boldsymbol{v}_i(0) = z_i(0) = v_0 \mathbb{1}_{|\mathcal{N}_i|h}$ and $\boldsymbol{\lambda}_i(0) = \boldsymbol{\xi}_i(0) = \lambda_0 \mathbb{1}_{|\mathcal{N}_i|h}$, for any $v_0, \lambda_0 \in \mathbb{R}$. Moreover, the initialization of the auxiliary variable $\boldsymbol{y}_i(0)$ in Algorithm 5.2 can also be carried out in a similar manner. □

Remark 5.2 Algorithms 3.1 and 3.2, which are performed under Assumption 3.1, are special cases of Algorithms 5.1 and 5.2, respectively, where $\beta_{pq} = 1$, for all $\{p, q\} \in \mathcal{E}^{\mathrm{c}}$ and $\gamma_p = 1$, for all $p \in \mathcal{P}$. □

5.4 Convergence Analysis

In this section, the convergence of the sequences produced by Algorithms 5.1 and 5.2 is studied. Moreover, the convergence rate, in terms of the ergodic average of the variables, is also provided.

5.4.1 Convergence of Stochastic DAL Algorithm

The main idea of the convergence analysis of Algorithm 5.1 is similar to that of Algorithm 3.1, which is shown in Sect. 3.2.2. In fact, the convergence analysis of Algorithm 5.1 is built upon the results obtained in the aforementioned section.

Algorithm 5.1 Stochastic distributed augmented Lagrangian (DAL) method

Initialization: For each node $i \in \mathcal{N}$, $\boldsymbol{v}_i(0) = \boldsymbol{v}_{i0} \in \mathbb{R}^{|\mathcal{N}_i|h}$ and $\boldsymbol{\lambda}_i(0) = \boldsymbol{\lambda}_{i0} \in \mathbb{R}^{|\mathcal{N}_i|h}$. Moreover, $z_i^j(0) = \boldsymbol{v}_j^i(0)$ and $\boldsymbol{\xi}_i^j(0) = \boldsymbol{\lambda}_j^i(0)$, for all $j \in \mathcal{N}_i$ and $i \in \mathcal{N}$.

Iteration: For each agent $p \in \mathcal{A}(k+1)$,

1. Update $(\boldsymbol{u}_i(k+1), \hat{\boldsymbol{v}}_i(k))$, for all $i \in \mathcal{M}_p$, according to

$$\{(\boldsymbol{u}_i(k+1), \hat{\boldsymbol{v}}_i(k))\}_{i \in \mathcal{M}_p} = \arg \min_{\{(\boldsymbol{u}_i, \boldsymbol{v}_i) \in \mathcal{L}_i\}_{i \in \mathcal{M}_p}} \sum_{i \in \mathcal{M}_p} \Bigg(f_i^{\mathrm{l}}(\boldsymbol{u}_i) + f_i^{\mathrm{c}}(\boldsymbol{v}_i) + $$
$$+ \sum_{j \in \mathcal{N}_i} \left(\langle \boldsymbol{\lambda}_i^j(k) + \boldsymbol{\xi}_i^j(k), \boldsymbol{v}_i^j \rangle + \| \boldsymbol{v}_i^j + z_i^j(k) \|_2^2 \right) \Bigg). \tag{5.1}$$

2. Update $\boldsymbol{v}_i^j(k+1)$, for all $j \in \mathcal{N}_i$ and $i \in \mathcal{M}_p$, as follows:

$$\boldsymbol{v}_i^j(k+1) = \begin{cases} \eta_i^j \hat{\boldsymbol{v}}_i^j(k) + \left(1 - \eta_i^j\right) \boldsymbol{v}_i^j(k), & \text{if } \phi(j, \cdot) \in \mathcal{A}_p(k+1) \cup \{p\}, \\ \boldsymbol{v}_i^j(k), & \text{otherwise.} \end{cases} \tag{5.2}$$

3. For each $j \in \mathcal{N}_i \backslash \mathcal{M}_p$ such that $\phi(j, \cdot) \in \mathcal{A}_p(k+1)$ and $i \in \mathcal{M}_p$, send $\boldsymbol{v}_i^j(k+1)$ to and receive $\boldsymbol{v}_j^i(k+1)$ from agent $\phi(j, \cdot)$.
4. Update the auxiliary and dual variables $z_i(k+1)$ and $\boldsymbol{\lambda}_i(k+1)$, for all $i \in \mathcal{M}_p$, according to

$$z_i^j(k+1) = \begin{cases} \boldsymbol{v}_j^i(k+1), & \text{if } \phi(j, \cdot) \in \mathcal{A}_p(k+1) \cup \{p\}, \\ z_i^j(k), & \text{otherwise,} \end{cases} \tag{5.3}$$

$$\boldsymbol{\lambda}_i^j(k+1) = \begin{cases} \boldsymbol{\lambda}_i^j(k) + \eta_i^j \left(\boldsymbol{v}_i^j(k+1) + z_i^j(k+1) \right), & \text{if } \phi(j, \cdot) \in \mathcal{A}_p(k+1) \cup \{p\}, \\ \boldsymbol{\lambda}_i^j(k), & \text{otherwise.} \end{cases} \tag{5.4}$$

5. For each $j \in \mathcal{N}_i \backslash \mathcal{M}_p$ such that $\phi(j, \cdot) \in \mathcal{A}_p(k+1)$ and $i \in \mathcal{M}_p$, send $\boldsymbol{\lambda}_i^j(k+1)$ to and receive $\boldsymbol{\lambda}_j^i(k+1)$ from agent $\phi(j, \cdot)$.
6. Update the auxiliary variable $\boldsymbol{\xi}_i(k+1)$, for all $i \in \mathcal{M}_p$, according to

$$\boldsymbol{\xi}_i^j(k+1) = \begin{cases} \boldsymbol{\lambda}_j^i(k+1), & \text{if } \phi(j, \cdot) \in \mathcal{A}_p(k+1) \cup \{p\}, \\ \boldsymbol{\xi}_i^j(k), & \text{otherwise.} \end{cases} \tag{5.5}$$

For agent $p \notin \mathcal{A}(k+1)$, $\boldsymbol{u}_i(k+1) = \boldsymbol{u}_i(k)$, $\boldsymbol{v}_i(k+1) = \boldsymbol{v}_i(k)$, $z_i(k+1) = z_i(k)$, $\boldsymbol{\lambda}_i(k+1) = \boldsymbol{\lambda}_i(k)$, and $\boldsymbol{\xi}_i(k+1) = \boldsymbol{\xi}_i(k)$, for all $i \in \mathcal{M}_p$.

The first step of this analysis is defining a Lyapunov function, which is denoted by $\tilde{V}_1(k)$ and constructed based on the Lyapunov function $V_1(k)$ defined in (3.19). For any saddle point of $L_1(\boldsymbol{u}, \boldsymbol{v}, \boldsymbol{\lambda})$ (see (3.3)), denoted by $(\boldsymbol{u}^\star, \boldsymbol{v}^\star, \boldsymbol{\lambda}^\star)$, $V_1(k)$ is defined as follows:

Algorithm 5.2 Stochastic distributed ADMM-based method

Initialization: For each $i \in \mathcal{N}$, $\boldsymbol{\mu}_i(0) = \boldsymbol{\mu}_{i0} \in \mathbb{R}^{|\mathcal{N}_i|h}$, and $\boldsymbol{y}_i^j(0) = \boldsymbol{y}_j^i(0) \in \mathbb{R}^h$, for all $j \in \mathcal{N}_i$.
Iteration: For each agent $p \in \mathcal{A}(k+1)$,

1. Update $\boldsymbol{u}_i(k+1)$ and $\boldsymbol{v}_i(k+1)$, for all $i \in \mathcal{M}_p$, according to

$$\begin{aligned}\{(\boldsymbol{u}_i(k+1), \boldsymbol{v}_i(k+1))\}_{i\in\mathcal{M}_p} = \arg \min_{\{(\boldsymbol{u}_i,\boldsymbol{v}_i)\in\mathcal{L}_i\}_{i\in\mathcal{M}_p}} \sum_{i\in\mathcal{M}_p} \Big(f_i^{\mathrm{l}}(\boldsymbol{u}_i) + f_i^{\mathrm{c}}(\boldsymbol{v}_i) \\ + \sum_{j\in\mathcal{N}_i} \langle \boldsymbol{\mu}_i^j(k), \boldsymbol{v}_i^j\rangle + \frac{1}{2}\|\boldsymbol{v}_i^j - \boldsymbol{y}_i^j(k)\|_2^2 \Big).\end{aligned} \tag{5.6}$$

2. For each $j \in \mathcal{N}_i \backslash \mathcal{M}_p$ such that $\phi(j,\cdot) \in \mathcal{A}_p(k+1)$ and $i \in \mathcal{M}_p$, send $\boldsymbol{v}_i^j(k+1)$ and $\boldsymbol{\mu}_i^j(k)$ to agent $\phi(j,\cdot) \in \mathcal{N}_p^{\mathrm{s}}$ and receive $\boldsymbol{v}_j^i(k+1)$ and $\boldsymbol{\mu}_j^i(k)$ from agent $\phi(j,\cdot) \in \mathcal{N}_p^{\mathrm{s}}$.
3. Update $\boldsymbol{y}_i^j(k+1)$, for all $j \in \mathcal{N}_i$ and $i \in \mathcal{M}_p$, as follows:

$$\boldsymbol{y}_i^j(k+1) = \begin{cases} \frac{1}{2}\left(\boldsymbol{\mu}_i^j(k) - \boldsymbol{\mu}_j^i(k) + \boldsymbol{v}_i^j(k+1) - \boldsymbol{v}_j^i(k+1)\right), \\ \qquad \text{if } \phi(j,\cdot) \in \mathcal{A}_p(k+1) \cup \{p\}, \\ \boldsymbol{y}_i^j(k), \quad \text{otherwise.} \end{cases} \tag{5.7}$$

4. Update the dual variables $\boldsymbol{\mu}_i^j(k+1)$, for all $j \in \mathcal{N}_i$ and $i \in \mathcal{M}_p$, according to

$$\boldsymbol{\mu}_i^j(k+1) = \begin{cases} \boldsymbol{\mu}_i^j(k) + \eta_i^j\left(\boldsymbol{v}_i^j(k+1) - \boldsymbol{y}_i^j(k+1)\right), & \text{if } \phi(j,\cdot) \in \mathcal{A}_p(k+1) \cup \{p\}, \\ \boldsymbol{\mu}_i^j(k), & \text{otherwise.} \end{cases} \tag{5.8}$$

For agent $p \notin \mathcal{A}(k+1)$, $\boldsymbol{u}_i(k+1) = \boldsymbol{u}_i(k)$, $\boldsymbol{v}_i(k+1) = \boldsymbol{v}_i(k)$, $\boldsymbol{y}_i(k+1) = \boldsymbol{y}_i(k)$, and $\boldsymbol{\mu}_i(k+1) = \boldsymbol{\mu}_i(k)$, for all $i \in \mathcal{M}_p$.

$$\tilde{V}_1(k) = \|\boldsymbol{v}(k) - \boldsymbol{v}^\star\|_{\tilde{H}}^2 + \frac{1}{2}\|\boldsymbol{\omega}(k) - \boldsymbol{\lambda}^\star\|_{\tilde{H}}^2, \tag{5.9}$$

where $\boldsymbol{\omega}(k) = [\boldsymbol{\omega}_i(k)]_{i\in\mathcal{N}}$, $\boldsymbol{\omega}_i(k) = [\boldsymbol{\omega}_i^j(k)]_{j\in\mathcal{N}_i}$, and

$$\boldsymbol{\omega}_i^j(k) = \boldsymbol{\lambda}_i^j(k) + (1 - \eta_i^j)(\boldsymbol{v}_i^j(k) + \boldsymbol{z}_i^j(k)), \ \forall j \in \mathcal{N}_i, \ \forall i \in \mathcal{N}.$$

Moreover, $\tilde{H} = \mathrm{blkdiag}(\{\tilde{H}_i\}_{i\in\mathcal{N}})$, $\tilde{H}_i = \mathrm{blkdiag}(\{(\alpha_{ij}\eta_i^j)^{-1} I_h\}_{j\in\mathcal{N}_i})$,for all $i \in \mathcal{N}$, where $\alpha_{ij} \in (0, 1]$, for each $j \in \mathcal{N}_i$ and $i \in \mathcal{N}$, is defined as follows. Let $p = \phi(i,\cdot) \in \mathcal{P}$ and $q = \phi(j,\cdot) \in \mathcal{P}$. Then,

$$\alpha_{ij} = \begin{cases} \beta_{pq}\gamma_p\gamma_q, & \text{if } p \neq q, \\ \gamma_p, & \text{otherwise.} \end{cases} \tag{5.10}$$

Recall that β_{pq} is the probability of communication link between agents p and q being active whereas γ_p is the probability of agent p being active, respectively (see Assumptions 5.1 and 5.2). In Lemma 5.1, it is shown that the sequence $\{\tilde{V}_1(k)\}$ is non-negative supermartingale. This fact is used as the basis of the convergence proof.

Lemma 5.1 *Let Assumptions 2.1–2.3 and 5.1–5.2 hold. Furthermore, let the sequence* $\{\boldsymbol{u}(k), \boldsymbol{v}(k), \hat{\boldsymbol{v}}(k), \boldsymbol{\lambda}(k)\}$ *be generated by Algorithm 5.1 and* $(\boldsymbol{u}^\star, \boldsymbol{v}^\star, \boldsymbol{\lambda}^\star)$ *be a saddle point of* $L_1(\boldsymbol{u}, \boldsymbol{v}, \boldsymbol{\lambda})$ *as defined in (3.3). Furthermore, consider* $\tilde{V}_1(k)$ *defined in (5.9). If* $\eta_i^j = \eta_j^i = \eta_{ij} \in \left(0, \frac{1}{4}\right)$, *for all* $j \in \mathcal{N}$ *and* $i \in \mathcal{N}$, *then the sequence* $\{\tilde{V}_1(k)\}$ *is a non-negative supermartingale and it holds with probability 1 that*

$$
\begin{aligned}
\mathbb{E}\left(\tilde{V}_1(k+1)|\mathcal{F}_1(k)\right) - \tilde{V}_1(k) \leq & -\sum_{i\in\mathcal{N}} \sigma_i \|\boldsymbol{u}_i(k+1) - \boldsymbol{u}_i^\star\|_2^2 \\
& -\sum_{i\in\mathcal{N}}\sum_{j\in\mathcal{N}_i} \left(\frac{3}{2} - \eta_{ij}\right) \|\hat{\boldsymbol{v}}_i^j(k) - \boldsymbol{v}_i^j(k)\|_2^2 \\
& -\sum_{i\in\mathcal{N}}\sum_{j\in\mathcal{N}_i} \frac{\eta_{ij} - (2\eta_{ij})^2}{2} \|\hat{\boldsymbol{v}}_i^j(k) + \hat{\boldsymbol{v}}_j^i(k)\|_2^2 \leq 0.
\end{aligned}
\tag{5.11}
$$

Proof Since $\tilde{V}_1(k)$ is a sum of norms and η_i^j and α_{ij} are positive, the sequence $\{\tilde{V}_1(k)\}$ is clearly non-negative. Denote by $\mathcal{F}_1(k)$ the filtration up to and including the iteration k, i.e., $\mathcal{F}_1(k) = \{\mathcal{A}(\ell), \mathcal{E}^c(\ell), \boldsymbol{u}(\ell), \boldsymbol{v}(\ell), \boldsymbol{\lambda}(\ell), \boldsymbol{z}(\ell), \boldsymbol{\xi}(\ell), \ \ell = 0, 1, \ldots, k\}$. Now, it must be shown that the conditional expectation of the sequence with respect to $\mathcal{F}_1(k)$ is always non-increasing. Based on Assumptions 5.1 and 5.2, the initialization step of Algorithm 5.1, and the update rules (5.2)–(5.4), the variables $\boldsymbol{v}_i^j(k+1), \boldsymbol{z}_i^j(k+1) = \boldsymbol{v}_j^i(k+1)$, and $\boldsymbol{\lambda}_i^j(k+1)$, for each $j \in \mathcal{N}_i \cap \mathcal{M}_p$ and $i \in \mathcal{M}_p$, are updated when agent $\phi(i, \cdot)$ is active, whereas, for each $j \in \mathcal{N}_i \backslash \mathcal{M}_p$ and $i \in \mathcal{M}_p$, those variables are only updated when agents $\phi(i, \cdot)$ and $\phi(j, \cdot)$ are active and communication link $\{\phi(i, \cdot), \phi(j, \cdot)\}$ is active. Therefore, it is observed from (5.10) that the probability of $\boldsymbol{v}_i^j(k+1), \boldsymbol{z}_i^j(k+1) = \boldsymbol{v}_j^i(k+1)$, and $\boldsymbol{\lambda}_i^j(k+1)$ being updated is $\alpha_{ij} \in (0, 1]$, whereas, with probability $1 - \alpha_{ij}$, they are not updated and the values remain the same as $\boldsymbol{v}_i^j(k)$, $\boldsymbol{z}_i^j(k) = \boldsymbol{v}_j^i(k)$, and $\boldsymbol{\lambda}_i^j(k)$. Thus, it is also observed that $\boldsymbol{\omega}_i^j(k+1) = \tilde{\boldsymbol{\lambda}}_i^j(k+1) = \boldsymbol{\lambda}_i^j(k+1) + (1 - \eta_i^j)(\boldsymbol{v}_i^j(k+1) + \boldsymbol{v}_j^i(k+1))$ with probability α_{ij} or the value $\boldsymbol{\omega}_i^j(k) = \tilde{\boldsymbol{\lambda}}_i^j(k)$ is kept with with probability $1 - \alpha_{ij}$. Hence, it is obtained with probability 1 that

$$\begin{aligned}
&\mathbb{E}\left(\tilde{V}_1(k+1)|\mathcal{F}_1(k)\right) - \tilde{V}_1(k)\\
&= \mathbb{E}\left(\|\boldsymbol{v}(k+1) - \boldsymbol{v}^\star\|_{\tilde{H}}^2 + \frac{1}{2}\|\boldsymbol{\omega}(k+1) - \boldsymbol{\lambda}^\star\|_{\tilde{H}}^2 \middle| \mathcal{F}_1(k)\right) - \tilde{V}_1(k)\\
&= \sum_{i\in\mathcal{N}}\sum_{j\in\mathcal{N}_i}\left(\frac{\alpha_{ij}}{\alpha_{ij}\eta_i^j}\|\boldsymbol{v}_i^j(k+1) - \boldsymbol{v}_i^{j\star}\|_2^2 + \frac{1-\alpha_{ij}}{\alpha_{ij}\eta_i^j}\|\boldsymbol{v}_i^j(k) - \boldsymbol{v}_i^{j\star}\|_2^2\right.\\
&\quad \left. + \frac{\alpha_{ij}}{2\alpha_{ij}\eta_i^j}\|\tilde{\boldsymbol{\lambda}}_i^j(k+1) - \boldsymbol{\lambda}_i^{j\star}\|_2^2 + \frac{1-\alpha_{ij}}{2\alpha_{ij}\eta_i^j}\|\tilde{\boldsymbol{\lambda}}_i^j(k) - \boldsymbol{\lambda}_i^{j\star}\|_2^2\right)\\
&\quad -\tilde{V}_1(k)\\
&= \|\boldsymbol{v}(k+1) - \boldsymbol{v}^\star\|_H^2 - \|\boldsymbol{v}(k) - \boldsymbol{v}^\star\|_H^2\\
&\quad + \frac{1}{2}\|\tilde{\boldsymbol{\lambda}}(k+1) - \boldsymbol{\lambda}^\star\|_H^2 - \frac{1}{2}\|\tilde{\boldsymbol{\lambda}}(k) - \boldsymbol{\lambda}^\star\|_H^2,
\end{aligned}$$

where the last equality is obtained since the terms $\sum_{i\in\mathcal{N}}\sum_{j\in\mathcal{N}_i}\frac{1}{\alpha_{ij}\eta_i^j}\|\boldsymbol{v}_i^j(k) - \boldsymbol{v}_i^{j\star}\|_2^2$, $\sum_{i\in\mathcal{N}}\sum_{j\in\mathcal{N}_i}\frac{1}{2\alpha_{ij}\eta_i^j}\|\tilde{\boldsymbol{\lambda}}_i^j(k) - \boldsymbol{\lambda}_i^{j\star}\|_2^2$, and $-\tilde{V}_1(k)$ cancel each other. Notice that since the scalings of the remaining quadratic terms do not involve α_{ij}, the weighted vector norm induced by H can be used. Based on the definition of $V_1(k)$ given in (3.19), it is obtained with probability 1 that

$$\mathbb{E}\left(\tilde{V}_1(k+1)|\mathcal{F}_1(k)\right) - \tilde{V}_1(k) = V_1(k+1) - V_1(k).$$

Therefore, by applying (3.20) to this relation, the desired relations in (5.11) follow, with probability 1, when $\eta_i^j = \eta_j^i = \eta_{ij} \in \left(0, \frac{1}{4}\right)$. Thus, (5.11) also shows that the sequence $\{\tilde{V}_1(k)\}$ is non-negative supermartingale. □

Now, the convergence theorem of Algorithm 5.1 is stated. The proof of the theorem relies on Lemma 5.1.

Theorem 5.1 *Let Assumptions 2.1–2.3 and 5.1–5.2 hold. Furthermore, let the sequence $\{\boldsymbol{u}(k), \boldsymbol{v}(k), \hat{\boldsymbol{v}}(k), \boldsymbol{\lambda}(k)\}$ be generated by Algorithm 5.1. If $\eta_i^j = \eta_j^i = \eta_{ij} \in \left(0, \frac{1}{4}\right)$, for all $j \in \mathcal{N}$ and $i \in \mathcal{N}$, then, with probability 1,*

a. (Feasibility) $\lim_{k\to\infty}\|\boldsymbol{v}_i^j(k) + \boldsymbol{v}_j^i(k)\|_2^2 = 0$, for all $j \in \mathcal{N}_i$ and $i \in \mathcal{N}$,
b. (Primal and dual variable convergence) There exists a saddle point of $L_1(\boldsymbol{u}, \boldsymbol{v}, \boldsymbol{\lambda})$ (see (3.3)), denoted by $(\boldsymbol{u}^\star, \boldsymbol{v}^\star, \boldsymbol{\lambda}^\star)$, such that $\lim_{k\to\infty}\boldsymbol{u}(k) = \boldsymbol{u}^\star$, $\lim_{k\to\infty}\boldsymbol{v}(k) = \boldsymbol{v}^\star$, and $\lim_{k\to\infty}\boldsymbol{\lambda}(k) = \boldsymbol{\lambda}^\star$. □

Proof Recall the function $\tilde{V}_1(k)$ defined in (5.9) and the inequality (5.11) in Lemma 5.1. Rearranging and iterating (5.11), for $\ell = 0, \ldots, k$, and taking the total expectation, it follows that

$$\sum_{\ell=0}^{k}\sum_{i\in\mathcal{N}}\left(\mathbb{E}\left(\sigma_i\|\boldsymbol{u}_i(\ell+1)-\boldsymbol{u}_i^\star\|_2^2\right)+\sum_{j\in\mathcal{N}_i}\left(\frac{3}{2}-\eta_{ij}\right)\mathbb{E}\left(\|\hat{\boldsymbol{v}}_i^j(\ell)-\boldsymbol{v}_i^j(\ell)\|_2^2\right)\right.$$
$$\left.+\sum_{j\in\mathcal{N}_i}\frac{\eta_{ij}-(2\eta_{ij})^2}{2}\mathbb{E}\left(\|\hat{\boldsymbol{v}}_i^j(\ell)+\hat{\boldsymbol{v}}_j^i(\ell)\|_2^2\right)\right)$$
$$\leq\sum_{\ell=0}^{k}\mathbb{E}\left(\tilde{V}_1(\ell)-\tilde{V}_1(\ell+1)\right)=\tilde{V}_1(0)-\mathbb{E}\left(\tilde{V}_1(k+1)\right)\leq\tilde{V}_1(0),$$

where the last inequality is obtained by dropping the non-positive term $-\mathbb{E}\left(\tilde{V}_1(k+1)\right)$. The above inequalities imply that $\{\mathbb{E}(\sigma_i\|\boldsymbol{u}_i(k+1)-\boldsymbol{u}_i^\star\|_2^2)\}$, for all $i\in\mathcal{N}$, is summable and converges to 0. Similarly, $\{\mathbb{E}(\|\hat{\boldsymbol{v}}_i^j(k)-\boldsymbol{v}_i^j(k)\|_2^2)\}$, and $\{\mathbb{E}(\|\hat{\boldsymbol{v}}_i^j(k)+\hat{\boldsymbol{v}}_j^i(k)\|_2^2)\}$, for all $j\in\mathcal{N}_i$ and $i\in\mathcal{N}$, are also summable and converge to 0. Using the Markov inequality, for any $\varepsilon>0$, it holds that

$$\limsup_{k\to\infty}\mathbb{P}\left(\sum_{i\in\mathcal{N}}\left(\sigma_i\|\boldsymbol{u}_i(k+1)-\boldsymbol{u}_i^\star\|_2^2+\sum_{j\in\mathcal{N}_i}\left(\frac{3}{2}-\eta_{ij}\right)\|\boldsymbol{v}_i^j(k)-\hat{\boldsymbol{v}}_i^j(k)\|_2^2\right)\right.$$
$$\left.+\sum_{i\in\mathcal{N}}\sum_{j\in\mathcal{N}_i}\frac{\eta_{ij}-(2\eta_{ij})^2}{2}\|\hat{\boldsymbol{v}}_i^j(k)+\hat{\boldsymbol{v}}_j^i(k)\|_2^2\geq\varepsilon\right)$$
$$\leq\limsup_{k\to\infty}\frac{1}{\varepsilon}\mathbb{E}\left(\sum_{i\in\mathcal{N}}\left(\sigma_i\|\boldsymbol{u}_i(k+1)-\boldsymbol{u}_i^\star\|_2^2+\sum_{j\in\mathcal{N}_i}\left(\frac{3}{2}-\eta_{ij}\right)\|\boldsymbol{v}_i^j(k)-\hat{\boldsymbol{v}}_i^j(k)\|_2^2\right)\right.$$
$$\left.+\sum_{i\in\mathcal{N}}\sum_{j\in\mathcal{N}_i}\frac{\eta_{ij}-(2\eta_{ij})^2}{2}\|\hat{\boldsymbol{v}}_i^j(k)+\hat{\boldsymbol{v}}_j^i(k)\|_2^2\right)=0.$$

Thus, it holds with probability 1 that

$$\lim_{k\to\infty}\|\boldsymbol{u}_i(k)-\boldsymbol{u}_i^\star\|_2^2=0,\ \forall i\in\mathcal{N}, \tag{5.12}$$
$$\lim_{k\to\infty}\|\boldsymbol{v}_i^j(k)-\hat{\boldsymbol{v}}_i^j(k)\|_2^2=0,\ \forall j\in\mathcal{N}_i,\ \forall i\in\mathcal{N}, \tag{5.13}$$
$$\lim_{k\to\infty}\|\hat{\boldsymbol{v}}_i^j(k)+\hat{\boldsymbol{v}}_j^i(k)\|_2^2=0,\ \forall j\in\mathcal{N}_i,\ \forall i\in\mathcal{N}. \tag{5.14}$$

Moreover, based on (5.13) and (5.14), it follows with probability 1 that

$$\lim_{k\to\infty}\|\boldsymbol{v}_i^j(k)+\boldsymbol{v}_j^i(k)\|_2^2=0,\ \forall j\in\mathcal{N}_i,\ \forall i\in\mathcal{N}. \tag{5.15}$$

The remaining steps of the proof are quite similar to the proof of Theorem 3.1. Based on (5.11) and the martingale convergence theorem, the sequences $\{\|\boldsymbol{v}(k)-\boldsymbol{v}^\star\|_{\tilde{H}}^2\}$ and $\{\|\boldsymbol{\omega}(k)-\boldsymbol{\lambda}^\star\|_{\tilde{H}}^2\}$ are bounded with probability 1, i.e., there exist accumulation points of the sequences $\{\boldsymbol{v}(k)\}$ and $\{\boldsymbol{\omega}(k)\}$. Furthermore, $\{\boldsymbol{\lambda}(k)\}$ is also bounded with probability 1 and has accumulation points due to the boundedness of $\{\boldsymbol{\omega}(k)\}$, the relation in (5.15), and the fact that $z_i^j(k) = \boldsymbol{v}_j^i(k)$, for each k, which follows from the initialization of $z_i^j(k)$ in Algorithm 5.1 and the update rules (5.2)–(5.3). Then, let $\{(\boldsymbol{v}(k_\ell), \boldsymbol{\lambda}(k_\ell))\}$ be a convergent subsequence and assume that $(\boldsymbol{v}^{\mathrm{a}}, \boldsymbol{\lambda}^{\mathrm{a}})$ is its limit point. Therefore, for each $j \in \mathcal{N}_i$ and $i \in \mathcal{N}$, $\lim_{\ell\to\infty} z_i^j(k_\ell) = \lim_{\ell\to\infty} \boldsymbol{v}_j^i(k_\ell) = \boldsymbol{v}_j^{i\mathrm{a}}$ and $\lim_{\ell\to\infty} \boldsymbol{\xi}_i^j(k_\ell) = \lim_{\ell\to\infty} \boldsymbol{\lambda}_j^i(k_\ell) = \boldsymbol{\lambda}_j^{i\mathrm{a}}$ with probability 1 due to the initialization of the variables in Algorithm 5.1 and the update rules (5.3) and (5.5).

Now, it must be shown that $(\boldsymbol{u}^\star, \boldsymbol{v}^{\mathrm{a}}, \boldsymbol{\lambda}^{\mathrm{a}})$ is a saddle point of $L_1(\boldsymbol{u}, \boldsymbol{v}, \boldsymbol{\lambda})$, i.e., $(\boldsymbol{u}^\star, \boldsymbol{v}^{\mathrm{a}}, \boldsymbol{\lambda}^{\mathrm{a}})$ satisfies the inequalities in (3.6). Based on (5.15), $\boldsymbol{v}_i^{j\mathrm{a}} + \boldsymbol{v}_j^{i\mathrm{a}} = \lim_{\ell\to\infty} (\boldsymbol{v}_i^j(k_\ell) + \boldsymbol{v}_j^i(k_\ell)) = 0$, with probability 1, for all $j \in \mathcal{N}_i$ and $i \in \mathcal{N}$. Thus, for any $\boldsymbol{\lambda} \in \mathbb{R}^{\sum_{i\in\mathcal{N}} h|\mathcal{N}_i|}$, $L_1(\boldsymbol{u}^\star, \boldsymbol{v}^{\mathrm{a}}, \boldsymbol{\lambda}) = L_1(\boldsymbol{u}^\star, \boldsymbol{v}^{\mathrm{a}}, \boldsymbol{\lambda}^{\mathrm{a}})$, satisfying the first inequality in (3.6). It remains to show the second inequality in (3.6). Consider the update step (5.1), for all $i \in \mathcal{N}$, i.e.,

$$
\begin{aligned}
(\boldsymbol{u}(k+1), \hat{\boldsymbol{v}}(k)) = \arg \min_{(\boldsymbol{u}_i,\boldsymbol{v}_i)\in\mathcal{L}_i, i\in\mathcal{N}} & \sum_{i\in\mathcal{N}} \Big(f_i^{\mathrm{l}}(\boldsymbol{u}_i) + f_i^{\mathrm{c}}(\boldsymbol{v}_i) \\
& + \sum_{j\in\mathcal{N}_i} \big(\langle \boldsymbol{\lambda}_i^j(k) + \boldsymbol{\xi}_i^j(k), \boldsymbol{v}_i^j \rangle + \|\boldsymbol{v}_i^j + z_i^j(k)\|_2^2 \big) \Big).
\end{aligned}
$$

By substituting k with k_ℓ and taking the limit as ℓ goes to infinity on both sides of the equality, it holds with probability 1 that

$$
\begin{aligned}
(\boldsymbol{u}^\star, \boldsymbol{v}^{\mathrm{a}}) = & \lim_{\ell\to\infty} \arg \min_{(\boldsymbol{u}_i,\boldsymbol{v}_i)\in\mathcal{L}_i, i\in\mathcal{N}} \sum_{i\in\mathcal{N}} \Big(f_i^{\mathrm{l}}(\boldsymbol{u}_i) + f_i^{\mathrm{c}}(\boldsymbol{v}_i) \\
& + \sum_{j\in\mathcal{N}_i} \big(\langle \boldsymbol{\lambda}_i^j(k_\ell) + \boldsymbol{\xi}_i^j(k_\ell), \boldsymbol{v}_i^j \rangle + \|\boldsymbol{v}_i^j + z_i^j(k_\ell)\|_2^2 \big) \Big) \\
= & \arg \min_{(\boldsymbol{u}_i,\boldsymbol{v}_i)\in\mathcal{L}_i, i\in\mathcal{N}} \sum_{i\in\mathcal{N}} \Big(f_i^{\mathrm{l}}(\boldsymbol{u}_i) + f_i^{\mathrm{c}}(\boldsymbol{v}_i) + \sum_{j\in\mathcal{N}_i} \big(\langle \boldsymbol{\lambda}_i^{j\mathrm{a}} + \boldsymbol{\lambda}_j^{i\mathrm{a}}, \boldsymbol{v}_i^j \rangle + \|\boldsymbol{v}_i^j + \boldsymbol{v}_j^{i\mathrm{a}}\|_2^2 \big) \Big) \\
= & \arg \min_{(\boldsymbol{u}_i,\boldsymbol{v}_i)\in\mathcal{L}_i, i\in\mathcal{N}} \sum_{i\in\mathcal{N}} \Big(f_i^{\mathrm{l}}(\boldsymbol{u}_i) + f_i^{\mathrm{c}}(\boldsymbol{v}_i) + \sum_{j\in\mathcal{N}_i} \langle \boldsymbol{\lambda}_i^{j\mathrm{a}}, \boldsymbol{v}_i^j + \boldsymbol{v}_j^i \rangle \Big).
\end{aligned}
\tag{5.16}
$$

The left-hand side of the first equality is obtained by using $\lim_{\ell\to\infty}(\boldsymbol{u}(k_\ell + 1), \hat{\boldsymbol{v}}(k_\ell)) = (\boldsymbol{u}^\star, \boldsymbol{v}^{\mathrm{a}})$, with probability 1, due to (5.12) and (5.13), which implies that $\lim_{\ell\to\infty} \hat{\boldsymbol{v}}(k_\ell) = \boldsymbol{v}^{\mathrm{a}}$, with probability 1. The second equality is obtained since $\lim_{\ell\to\infty} z_i^j(k_\ell) = \boldsymbol{v}_j^{i\mathrm{a}}$ and $\lim_{\ell\to\infty} \boldsymbol{\xi}_i^j(k_\ell) = \boldsymbol{\lambda}_j^{i\mathrm{a}}$, with probability 1, for all $j \in \mathcal{N}_i$ and $i \in \mathcal{N}$. Then, the last equality holds since the term $\sum_{i\in\mathcal{N}} \sum_{j\in\mathcal{N}_i} \|\boldsymbol{v}_i^j + \boldsymbol{v}_j^{i\mathrm{a}}\|_2^2$ is zero at $(\boldsymbol{u}^\star, \boldsymbol{v}^{\mathrm{a}})$ due to the fact that $\boldsymbol{v}_i^{j\mathrm{a}} + \boldsymbol{v}_j^{i\mathrm{a}} = 0$, for all $j \in \mathcal{N}_i$ and $i \in \mathcal{N}$. Additionally, $\boldsymbol{v}^{\mathrm{a}}$ is also an attainer of $\min_{\boldsymbol{v}} \sum_{i\in\mathcal{N}} \sum_{j\in\mathcal{N}_i} \|\boldsymbol{v}_i^j + \boldsymbol{v}_j^i\|_2^2$ since $\boldsymbol{v}_i^{j\mathrm{a}} + \boldsymbol{v}_j^{i\mathrm{a}} = 0$, for all $j \in \mathcal{N}_i$ and $i \in \mathcal{N}$. Therefore, the pair $(\boldsymbol{u}^\star, \boldsymbol{v}^{\mathrm{a}})$ also minimizes $L_1(\boldsymbol{u}, \boldsymbol{v}, \boldsymbol{\lambda}^{\mathrm{a}})$, i.e.,

$$(\boldsymbol{u}^\star, \boldsymbol{v}^{\mathrm{a}}) \in \arg \min_{(\boldsymbol{u}_i,\boldsymbol{v}_i)\in\mathcal{L}_i, i\in\mathcal{N}} \sum_{i\in\mathcal{N}} \Big(f_i^{\mathrm{l}}(\boldsymbol{u}_i) + f_i^{\mathrm{c}}(\boldsymbol{v}_i) + \sum_{j\in\mathcal{N}_i} \Big(\langle \boldsymbol{\lambda}_i^{j\mathrm{a}}, \boldsymbol{v}_i^j + \boldsymbol{v}_j^i \rangle + \|\boldsymbol{v}_i^j + \boldsymbol{v}_j^i\|_2^2 \Big) \Big),$$

where the cost function in the minimization is obtained by adding the quadratic term $\sum_{i\in\mathcal{N}} \sum_{j\in\mathcal{N}_i} \|\boldsymbol{v}_i^j + \boldsymbol{v}_j^i\|_2^2$ to the cost function on the right-hand side of the last equality in (5.16). Hence, the preceding relation implies the second inequality in (3.6). Thus, it is concluded that $(\boldsymbol{u}^\star, \boldsymbol{v}^{\mathrm{a}}, \boldsymbol{\lambda}^{\mathrm{a}})$ is a saddle point of $L_1(\boldsymbol{u}, \boldsymbol{v}, \boldsymbol{\lambda})$ with probability 1. Finally, set $\boldsymbol{v}^\star = \boldsymbol{v}^{\mathrm{a}}$ and $\boldsymbol{\lambda}^\star = \boldsymbol{\lambda}^{\mathrm{a}}$ in $\tilde{V}_1(k)$ (see (5.9)). Since the subsequence of $\tilde{V}_1(k_\ell)$ converges to 0 with probability 1 and $\tilde{V}_1(k)$ is non-negative supermartingale, the entire sequence $\{(\boldsymbol{v}(k), \boldsymbol{\lambda}(k)\}$ converges to $(\boldsymbol{v}^{\mathrm{a}}, \boldsymbol{\lambda}^{\mathrm{a}})$ with probability 1. □

Remark 5.3 If the dual variables $\boldsymbol{\lambda}_i(0)$, for all $i \in \mathcal{N}$, are initialized such that $\boldsymbol{\lambda}_{i0}^j = \boldsymbol{\lambda}_{j0}^i$, then, $\boldsymbol{\lambda}_i^j(k) = \boldsymbol{\lambda}_j^i(k)$, for all $k \geq 0$, since $\eta_i^j = \eta_j^i = \eta_{ij}$. In this setup, similarly as in Algorithm 3.1, the second round of communication (Step 5) in Algorithm 5.1 is not necessary and each agent $p \in \mathcal{A}(k+1)$ can update $\boldsymbol{\xi}_i^j(k+1) = \boldsymbol{\lambda}_i^j(k+1)$, for each $j \in \mathcal{N}_i \backslash \mathcal{M}_p$ and $i \in \mathcal{M}_p$, if $\phi(j, \cdot) \in \mathcal{A}_p(k+1) \cup \{p\}$, or $\boldsymbol{\xi}_i^j(k+1) = \boldsymbol{\xi}_i^j(k)$, otherwise. □

Next, the convergence rate of Algorithm 5.1 is shown. To that end, define the ergodic average of the primal and auxiliary variables, as follows:

$$\bar{\boldsymbol{u}}_i(k) = \sum_{\ell=0}^{k-1} \frac{\boldsymbol{u}_i(\ell)}{k}, \; \bar{\boldsymbol{v}}_i(k) = \sum_{\ell=0}^{k-1} \frac{\boldsymbol{v}_i(\ell)}{k}, \; \bar{\hat{\boldsymbol{v}}}_i(k) = \sum_{\ell=0}^{k-1} \frac{\hat{\boldsymbol{v}}_i(\ell)}{k}, \tag{5.17}$$

for all $i \in \mathcal{N}$.

Theorem 5.2 *Let Assumptions 2.1–2.3 and 5.1–5.2 hold. Furthermore, let the sequence* $\{\boldsymbol{u}(k), \boldsymbol{v}(k), \hat{\boldsymbol{v}}(k), \boldsymbol{\lambda}(k)\}$ *be generated by Algorithm 5.1 with* $\eta_i^j = \eta_j^i = \eta_{ij} \in \left(0, \frac{1}{4}\right)$, *for all* $j \in \mathcal{N}$ *and* $i \in \mathcal{N}$, *and* $(\boldsymbol{u}^\star, \boldsymbol{v}^\star, \boldsymbol{\lambda}^\star)$ *be a saddle point of* $L_1(\boldsymbol{u}, \boldsymbol{v}, \boldsymbol{\lambda})$ *as defined in (3.3). Then, the convergence rate of Algorithm 5.1 is* $\mathcal{O}(\frac{1}{k})$ *and it holds that*

$$\begin{aligned}
&\sum_{i\in\mathcal{N}} \mathbb{E}\left(\sigma_i \|\bar{\boldsymbol{u}}_i(k) - \boldsymbol{u}_i^\star\|_2^2\right) + \sum_{i\in\mathcal{N}} \sum_{j\in\mathcal{N}_i} \left(\frac{3}{2} - \eta_{ij}\right) \mathbb{E}\left(\|\bar{\hat{\boldsymbol{v}}}_i^j(k-1) - \bar{\boldsymbol{v}}_i^j(k-1)\|_2^2\right) \\
&+ \sum_{i\in\mathcal{N}} \sum_{j\in\mathcal{N}_i} \frac{\eta_{ij} - (2\eta_{ij})^2}{2} \mathbb{E}\left(\|\bar{\hat{\boldsymbol{v}}}_i^j(k-1) + \bar{\hat{\boldsymbol{v}}}_j^i(k-1)\|_2^2\right) \\
&\leq \frac{1}{\underline{\alpha} k}\left(\|\boldsymbol{v}(0) - \boldsymbol{v}^\star\|_H^2 + \frac{1}{2}\|\tilde{\boldsymbol{\lambda}}(0) - \boldsymbol{\lambda}^\star\|_H^2\right),
\end{aligned} \tag{5.18}$$

where $\bar{\boldsymbol{u}}_i(k)$, $\bar{\boldsymbol{v}}_i(k)$, *and* $\bar{\hat{\boldsymbol{v}}}_i(k)$ *are the ergodic average of the primal variables (5.17),* $\underline{\alpha} = \min_{\{i,j\}\in\mathcal{E}} \alpha_{ij}$, *and* $\tilde{\boldsymbol{\lambda}}(k)$ *is defined by (3.17).* □

Proof By rearranging the summation of (5.11) over $\ell = 0, \ldots, k-1$ and taking the total expectation, it holds that

$$\sum_{\ell=0}^{k-1}\sum_{i\in\mathcal{N}}\left(\mathbb{E}\left(\sigma_i\|\boldsymbol{u}_i(\ell+1)-\boldsymbol{u}_i^\star\|_2^2\right)+\sum_{j\in\mathcal{N}_i}\left(\frac{3}{2}-\eta_{ij}\right)\mathbb{E}\left(\|\hat{\boldsymbol{v}}_i^j(\ell)-\boldsymbol{v}_i^j(\ell)\|_2^2\right)\right.$$
$$\left.+\sum_{j\in\mathcal{N}_i}\frac{\eta_{ij}-(2\eta_{ij})^2}{2}\mathbb{E}\left(\|\hat{\boldsymbol{v}}_i^j(\ell)+\hat{\boldsymbol{v}}_j^i(\ell)\|_2^2\right)\right)\leq\sum_{\ell=0}^{k-1}\mathbb{E}\left(\tilde{V}_1(\ell)-\tilde{V}_1(\ell+1)\right) \tag{5.19}$$
$$=\tilde{V}_1(0)-\mathbb{E}\left(\tilde{V}_1(k)\right)\leq\tilde{V}_1(0)=\|\boldsymbol{v}(0)-\boldsymbol{v}^\star\|_{\tilde{H}}^2+\frac{1}{2}\|\tilde{\boldsymbol{\lambda}}(0)-\boldsymbol{\lambda}^\star\|_{\tilde{H}}^2$$
$$\leq\frac{1}{\underline{\alpha}}\|\boldsymbol{v}(0)-\boldsymbol{v}^\star\|_H^2+\frac{1}{2\underline{\alpha}}\|\tilde{\boldsymbol{\lambda}}(0)-\boldsymbol{\lambda}^\star\|_H^2$$

where the second inequality is obtained by dropping the non-positive term $-\mathbb{E}\left(\tilde{V}_1(k)\right)$ and the last inequality is obtained by using the definition of $\underline{\alpha}$. Furthermore, due to the convexity of the squared of the Euclidean norm, it follows that

$$k\mathbb{E}(\|\bar{\boldsymbol{u}}_i(k)-\boldsymbol{u}_i^\star\|_2^2)\leq\sum_{\ell=0}^{k-1}\mathbb{E}(\|\boldsymbol{u}_i(\ell+1)-\boldsymbol{u}_i^\star\|_2^2),$$
$$k\mathbb{E}(\|\bar{\hat{\boldsymbol{v}}}_i^j(k-1)-\bar{\boldsymbol{v}}_i^j(k-1)\|_2^2)\leq\sum_{\ell=0}^{k-1}\mathbb{E}(\|\hat{\boldsymbol{v}}_i^j(\ell)-\boldsymbol{v}_i^j(\ell)\|_2^2),$$
$$k\mathbb{E}(\|\bar{\hat{\boldsymbol{v}}}_i^j(k-1)+\bar{\hat{\boldsymbol{v}}}_j^i(k-1)\|_2^2)\leq\sum_{\ell=0}^{k-1}\mathbb{E}(\|\hat{\boldsymbol{v}}_i^j(\ell)+\hat{\boldsymbol{v}}_j^i(\ell)\|_2^2).$$

The desired inequality follows by applying the above relations to (5.19) and using the fact that $\sigma_i>0$, for all $i\in\mathcal{N}$, and $\frac{3}{2}-\eta_{ij}>0$, $\frac{\eta_{ij}-(2\eta_{ij})^2}{2}>0$, for all $j\in\mathcal{N}_i$ and $i\in\mathcal{N}$. □

Theorem 5.2 shows that the convergence rate of Algorithm 5.1 in terms of the ergodic average of the primal and dual variables is $\mathcal{O}(\frac{1}{k})$. Furthermore, it also implies that the convergence time is $\mathcal{O}(\frac{1}{\epsilon\underline{\alpha}})$, showing the influence of the probability of the activation of links and agents on the number of iterations required to reach a solution with the desired error ϵ.

5.4.2 Convergence of Stochastic Distributed ADMM Algorithm

The convergence of the sequence produced by Algorithm 5.2 is shown using the same technique used in Sect. 5.4.1. First, a suitable Lyapunov function is defined. To that end, recall α_{ij} from (5.10) and $\tilde{H}=\text{blkdiag}(\{\tilde{H}_i\}_{i\in\mathcal{N}})$ and $\tilde{H}_i=\text{blkdiag}(\{(\alpha_{ij}\eta_i^j)^{-1}I_h\}_{j\in\mathcal{N}_i})$, for all $i\in\mathcal{N}$. Moreover, define that $G=\text{blkdiag}(\{G_i\}_{i\in\mathcal{N}})$ and

$G_i = \text{blkdiag}(\{(\alpha_{ij})^{-1} I_h\}_{j \in \mathcal{N}_i})$, for all $i \in \mathcal{N}$. Then, the candidate Lyapunov function, for any saddle point of $L_2(\boldsymbol{u}, \boldsymbol{v}, \boldsymbol{y}, \boldsymbol{\mu})$ (see (3.29)), which is denoted by $(\boldsymbol{u}^\star, \boldsymbol{v}^\star, \boldsymbol{y}^\star, \boldsymbol{\mu}^\star)$, is defined by

$$\tilde{V}_2(k) = \frac{1}{2}\|\boldsymbol{\mu}(k) - \boldsymbol{\mu}^\star\|_{\tilde{H}} + \frac{1}{2}\|\boldsymbol{y}(k) - \boldsymbol{y}^\star\|_G. \tag{5.20}$$

Lemma 5.2 shows that the sequence $\{V_2(k)\}$ is non-negative supermartingale.

Lemma 5.2 *Let Assumptions 2.1–2.3 and 5.1–5.2 hold. Furthermore, let the sequence $\{\boldsymbol{u}(k), \boldsymbol{v}(k), \boldsymbol{y}(k), \boldsymbol{\mu}(k)\}$ be generated by Algorithm 5.2 and $(\boldsymbol{u}^\star, \boldsymbol{v}^\star, \boldsymbol{y}^\star, \boldsymbol{\mu}^\star)$ be a saddle point of $L_2(\boldsymbol{u}, \boldsymbol{v}, \boldsymbol{y}, \boldsymbol{\mu})$ as defined in (3.29). Furthermore, consider $\tilde{V}_2(k)$ defined in (5.20). If $\eta_i^j \in (0, 1)$, for all $j \in \mathcal{N}$ and $i \in \mathcal{N}$, then the sequence $\{\tilde{V}_2(k)\}$ is a non-negative supermartingale and it holds with probability 1 that*

$$\begin{aligned}\mathbb{E}\left(\tilde{V}_2(k+1)|\mathcal{F}_1(k)\right) - \tilde{V}_2(k) \le &-\frac{1}{2}\|\boldsymbol{r}(k+1) + \boldsymbol{y}(k+1) - \boldsymbol{y}(k)\|_2^2 \\ &- \sum_{i \in \mathcal{N}} \sigma_i \|\boldsymbol{u}_i(k+1) - \boldsymbol{u}_i^\star\|_2^2 \\ &- \sum_{i \in \mathcal{N}} \sum_{j \in \mathcal{N}_i} \frac{1 - \eta_i^j}{2} \|\boldsymbol{r}_i^j(k+1)\|_2^2 \le 0.\end{aligned} \tag{5.21}$$

Proof The proof follows the same steps in the proof of Lemma 5.1. For completeness, the proof is shown in Appendix A.2. □

Now, the convergence properties of Algorithm 5.2 are stated as follows.

Theorem 5.3 *Let Assumptions 2.1–2.3 and 5.1–5.2 hold. Furthermore, let the sequence $\{\boldsymbol{u}(k), \boldsymbol{v}(k), \boldsymbol{y}(k), \boldsymbol{\mu}(k)\}$ be generated by Algorithm 5.2. If $\eta_i^j \in (0, 1)$, for all $j \in \mathcal{N}$ and $i \in \mathcal{N}$, then,*

a. (Feasibility) $\lim_{k\to\infty} \|\boldsymbol{v}_i^j(k) + \boldsymbol{v}_j^i(k)\|_2^2 = 0$, with probability 1, for all $j \in \mathcal{N}_i$ and $i \in \mathcal{N}$.
b. (Primal and dual variable convergence) There exists a saddle point of $L_2(\boldsymbol{u}, \boldsymbol{v}, \boldsymbol{y}, \boldsymbol{\mu})$ (see (3.29)), denoted by $(\boldsymbol{u}^\star, \boldsymbol{v}^\star, \boldsymbol{y}^\star, \boldsymbol{\mu}^\star)$, such that $\lim_{k\to\infty} \boldsymbol{u}(k) = \boldsymbol{u}^\star$, $\lim_{k\to\infty} \boldsymbol{v}(k) = \boldsymbol{v}^\star$, $\lim_{k\to\infty} \boldsymbol{y}(k) = \boldsymbol{y}^\star$, and $\lim_{k\to\infty} \boldsymbol{\mu}(k) = \boldsymbol{\mu}^\star$, with probability 1. □

Proof The steps of the proof follows that of Theorem 5.1. By rearranging and iterating the first inequality in (5.21) over $\ell = 0, 1, \ldots, k$ and taking the total expectation, it follows that

$$\lim_{k\to\infty} \|\boldsymbol{r}(k+1) + \boldsymbol{y}(k+1) - \boldsymbol{y}(k)\|_2^2 = 0, \tag{5.22}$$

$$\lim_{k\to\infty} \sigma_i \|\boldsymbol{u}_i(k+1) - \boldsymbol{u}_i^\star\|_2^2 = 0, \quad \forall i \in \mathcal{N}, \tag{5.23}$$

$$\lim_{k\to\infty} \|\boldsymbol{r}(k+1)\|_2^2 = \lim_{k\to\infty} \|\boldsymbol{v}(k+1) - \boldsymbol{y}(k+1)\|_2^2 = 0, \tag{5.24}$$

with probability 1. Moreover, by (5.22) and (5.24), it holds with probability 1 that

$$\lim_{k\to\infty} \|\boldsymbol{y}(k+1) - \boldsymbol{y}(k)\|_2^2 = 0, \tag{5.25}$$

with probability 1. Additionally, due to the initialization step in Algorithm 5.2 and the update rule in (5.7), $\boldsymbol{y}_i^j(k) = -\boldsymbol{y}_j^i(k)$, for all $j \in \mathcal{N}_i$, $i \in \mathcal{N}$, and $k \in \mathbb{Z}_{\geq 0}$. Based on this fact and the relation in (5.24), it also holds that

$$\lim_{k\to\infty} \|\boldsymbol{v}_i^j(k+1) + \boldsymbol{v}_j^i(k+1)\|_2^2 = 0, \;\forall j \in \mathcal{N}_i, \;\forall i \in \mathcal{N}. \tag{5.26}$$

Then, it remains to show that the sequence $\{(\boldsymbol{u}(k), \boldsymbol{y}(k), \boldsymbol{\mu}(k))\}$ converges to a saddle point with probability 1. The arguments are similar to those that are in the proof of Theorem 3.2. The complete proof is given in Appendix A.3. □

Finally, the convergence rate of Algorithm 5.2, with respect to the ergodic average of the primal and auxiliary variables defined by

$$\bar{\boldsymbol{u}}(k) = \sum_{\ell=0}^{k-1} \frac{\boldsymbol{u}(\ell)}{k}, \; \bar{\boldsymbol{v}}(k) = \sum_{\ell=0}^{k-1} \frac{\boldsymbol{v}(\ell)}{k}, \; \bar{\boldsymbol{y}}(k) = \sum_{\ell=0}^{k-1} \frac{\hat{\boldsymbol{v}}(\ell)}{k}, \tag{5.27}$$

is stated as follows.

Theorem 5.4 *Let Assumptions 2.1–2.3 and 5.1–5.2 hold. Furthermore, let the sequence* $\{\boldsymbol{u}(k), \boldsymbol{v}(k), \boldsymbol{y}(k), \boldsymbol{\mu}(k)\}$ *be generated by Algorithm 5.2 with* $\eta_i^j \in (0, 1)$, *for all* $j \in \mathcal{N}$ *and* $i \in \mathcal{N}$, *and* $(\boldsymbol{u}^\star, \boldsymbol{v}^\star, \boldsymbol{\lambda}^\star)$ *be a saddle point of* $L_2(\boldsymbol{u}, \boldsymbol{v}, \boldsymbol{y}, \boldsymbol{\lambda})$ *as defined in (3.29). Then, the convergence rate of Algorithm 5.2 is* $\mathcal{O}(\frac{1}{k})$ *and it holds that*

$$\begin{aligned} &\frac{1}{2}\mathbb{E}\left(\|\bar{\boldsymbol{v}}(k) - \bar{\boldsymbol{y}}(k-1)\|_2^2\right) + \sum_{i\in\mathcal{N}} \mathbb{E}\left(\sigma_i \|\bar{\boldsymbol{u}}_i(k) - \boldsymbol{u}_i^\star\|_2^2\right) \\ &+ \sum_{i\in\mathcal{N}} \sum_{j\in\mathcal{N}_i} \frac{1-\eta_i^j}{2} \mathbb{E}\left(\|\bar{\boldsymbol{v}}_i^j(k) - \bar{\boldsymbol{y}}_i^j(k)\|_2^2\right) \\ &\leq \frac{1}{\underline{\alpha} k}\left(\frac{1}{2}\|\boldsymbol{y}(0) - \boldsymbol{y}^\star\|_2^2 + \frac{1}{2}\|\boldsymbol{\mu}(0) - \boldsymbol{\mu}^\star\|_H^2\right), \end{aligned} \tag{5.28}$$

where $\bar{\boldsymbol{u}}_i(k)$, $\bar{\boldsymbol{v}}_i(k)$, *and* $\bar{\boldsymbol{y}}_i(k)$ *are the ergodic average of the primal variables (5.27) and* $\underline{\alpha} = \min_{\{i,j\}\in\mathcal{E}} \alpha_{ij}$. □

Proof The proof is similar to that of Theorem 5.2. For completeness, it is given in Appendix A.4. □

5.5 Numerical Simulations

This section showcases the performance of Algorithms 5.1 and 5.2 in a small-scale energy management problem. The network considered in this numerical study is depicted in Fig. 5.1. It is a network of prosumers where each node represents a prosumer as an agent. Moreover, each agent i has a dispatchable generator. It is considered that the dispatchable generator has a quadratic cost function, i.e.,

$$f_i^{\mathrm{l}}(\boldsymbol{u}_i) = c_i^{\mathrm{dg}} \boldsymbol{u}_i^\top \boldsymbol{u}_i,$$

where c_i^{dg} is a positive scalar that denotes the per-unit cost of generating power. It is also considered that the trading of power between agents is subject to a linear cost, i.e.,

$$f_i^{\mathrm{c}}(\boldsymbol{v}_i) = \sum_{j \in \mathcal{N}_i} c_{ij} \mathbb{1}^\top \boldsymbol{v}_i^j,$$

where c_{ij} is a positive scalar that denotes the per-unit cost of transfering power.

The parameter values of each agent are given in Table 5.1 and their loads are shown in Fig. 5.2. Moreover, Algorithm 5.1 uses $\eta_i^j = 0.2$, for all $j \in \mathcal{N}_i$ and $i \in \mathcal{N}$ whereas Algorithm 5.1 uses $\eta_i^j = 0.6$, for all $j \in \mathcal{N}_i$ and $i \in \mathcal{N}$. In these simulations, a one-shot optimization at the 18th hour with the time horizon 4 steps ahead is performed. Furthermore, it is assumed that $\beta_{ij} = \beta$, for all $\{i, j\} \in \mathcal{E}$ and $\gamma_i = \gamma$. Furthermore,

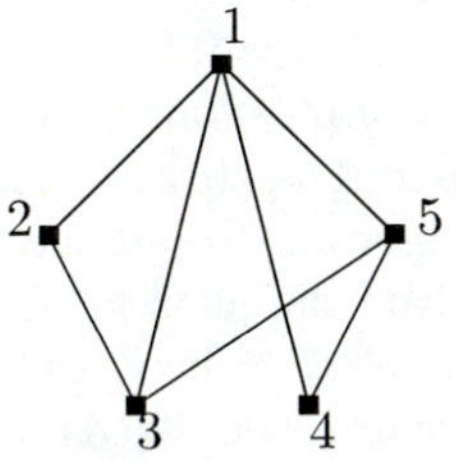

Fig. 5.1 The network of 5 agents. Squares (■) indicate the agents whereas the solid lines indicate the physical links between agents

Table 5.1 Parameter values of the agents

Parameters	Agent				
	1	2	3	4	5
c_i^{dg}	10	8	6	4	2
c_{ij}	1	1	1	1	1
$u_i^{\max}$	10	10	10	10	10
$v_i^{\max}$	3.5	3.5	3.5	3.5	3.5

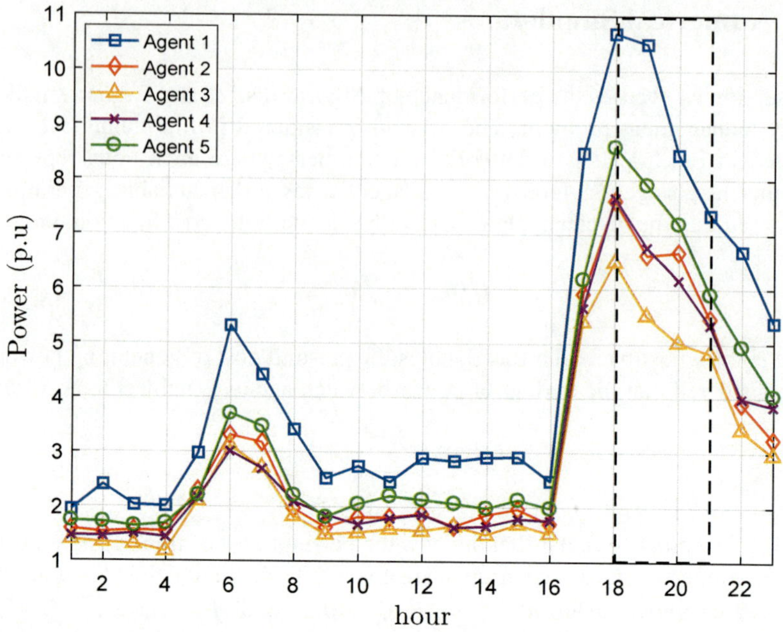

Fig. 5.2 One-day load data used in the simulations is obtained based on a residential load profile. The one-shot optimization simulated occurs at the 18th hour with load data used appear inside the dashed rectangle

four different values of α, i.e., $\alpha = 1, \alpha = 0.8, \alpha = 0.6$, and $\alpha = 0.4$, are considered. For each $\alpha < 1$, ten Monte Carlo simulations are carried out.

The convergences of $\|\boldsymbol{u}(k) - \boldsymbol{u}^\star\|_2$, $\|\boldsymbol{v}(k) - \hat{\boldsymbol{v}}(k)\|_2$, and $\|\boldsymbol{v}(k) + \boldsymbol{z}(k)\|_2$ with different values of α obtained by Algorithm 5.1 are shown in Fig. 5.3. On the other hand, the simulation results obtained from simulating Algorithm 5.2, i.e., the convergence of $\|\boldsymbol{u}(k) - \boldsymbol{u}^\star\|_2$ and $\|\boldsymbol{v}(k) - \boldsymbol{y}(k)\|_2$, are shown in Fig. 5.4. Note that the value $u^\star$ is computed by solving the problem in a centralized fashion, which can still be done due to the small size of the system. Furthermore, the optimal value of the problem, which is 4.91×10^3, is always obtained in all simulations. Moreover, it can be seen that the larger α, the faster the primal and dual variables converge under both algorithms, as shown in Theorems 5.2 and 5.4. In addition, it is worth mentioning that despite Algorithm 5.2 uses a larger step size than Algorithm 5.1, the former method reaches the same error value of $\boldsymbol{u}(k)$, i.e., $\|\boldsymbol{u}(k) - \boldsymbol{u}^\star\|_2$, slower than the latter method.

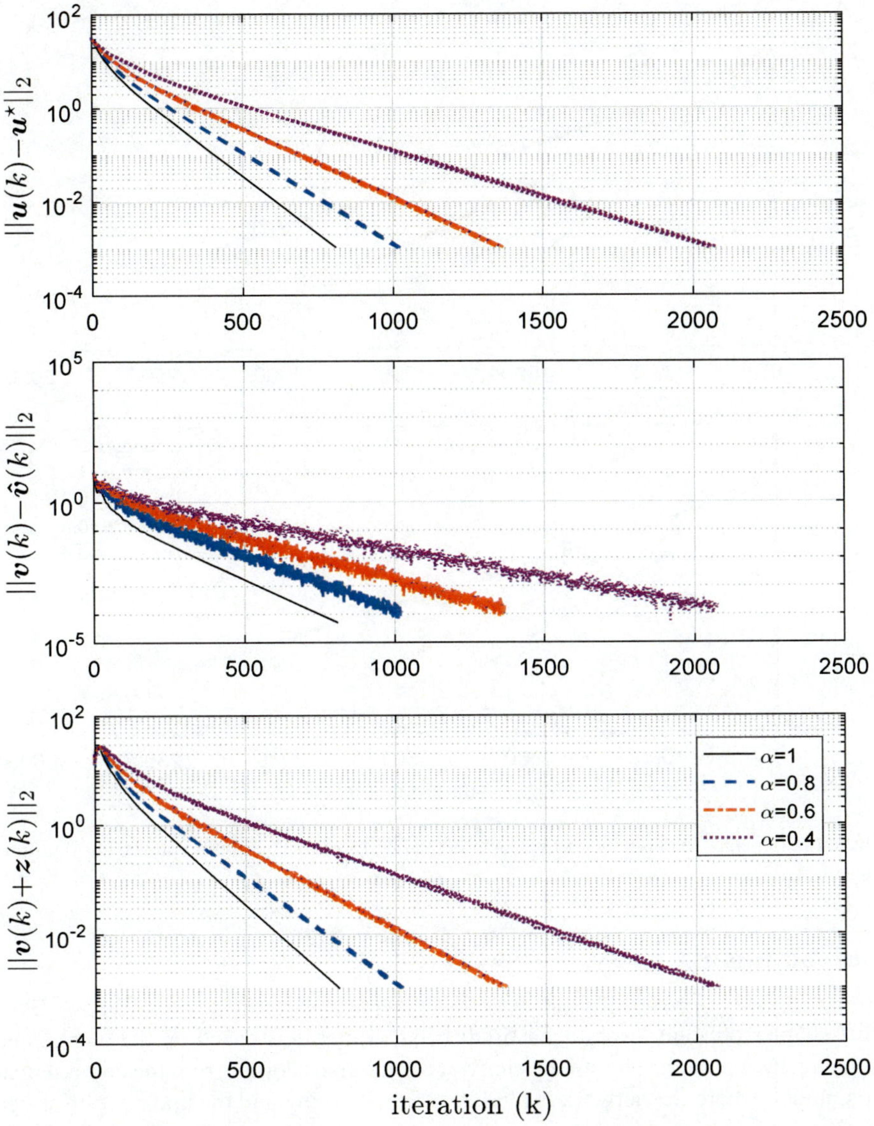

Fig. 5.3 Convergence of $\|u(k) - u^\star\|_2$ (top plot), $\|v(k) - \hat{v}(k)\|_2$ (middle plot) and $\|v(k) + z(k)\|_2$ (bottom plot) with Algorithm 5.1

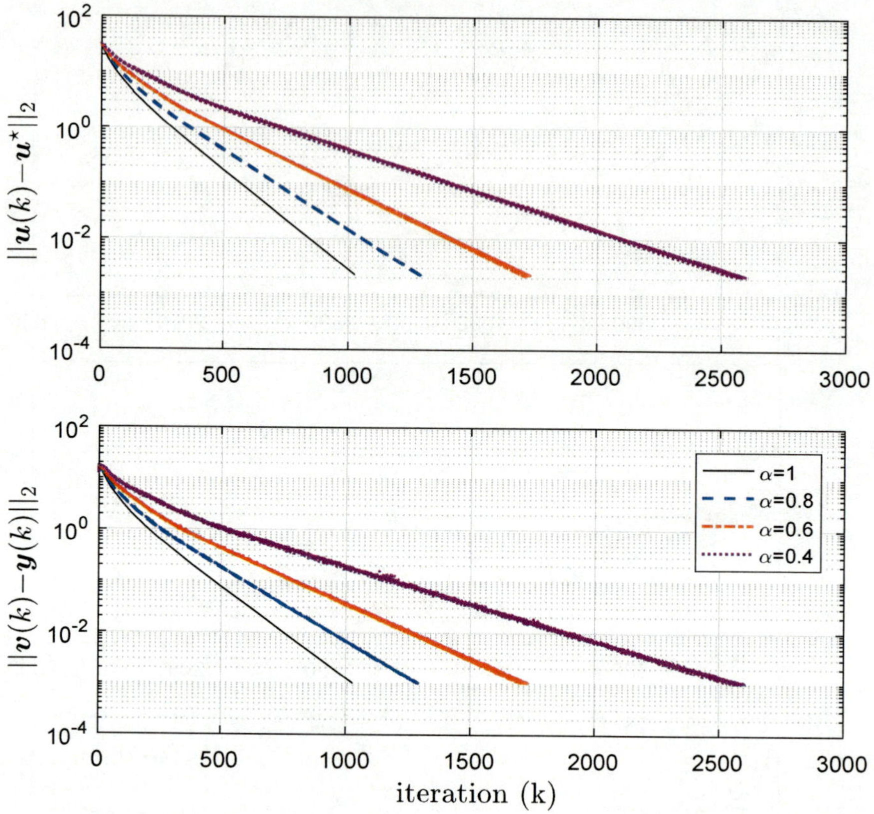

Fig. 5.4 Convergence of $\|\boldsymbol{u}(k) - \boldsymbol{u}^\star\|_2$ (top plot) and $\|\boldsymbol{v}(k) - \boldsymbol{y}(k)\|_2$ (bottom plot) with Algorithm 5.2

5.6 Summary

This chapter presents two stochastic distributed methods based on the augmented Lagrangian approach. The stochasticity of the methods comes from the communication model, where the network is randomly time-varying and the updates performed by the agents are randomly asynchronous. The convergence of the sequences produced by the proposed algorithms is analytically shown. Moreover, a numerical simulation with a small-scale network of prosumers shows their performance. Next, in Chap. 6, both algorithms are implemented as a distributed MPC scheme for economic dispatch. Some heuristic methods to increase the speed of convergence are introduced and the schemes are implemented in the benchmark case by means of simulations.

References

1. Larsen GKH, van Foreest ND, Scherpen JMA (2014) Distributed MPC applied to a network of households with micro-CHP and heat storage. IEEE Trans Smart Grid 5(4):2106–2114
2. Kraning M, Chu E, Lavaei J, Boyd S (2014) Dynamic network energy management via proximal message passing. Found Trends Optim 1(2):73–126
3. Wang T, O'Neill D, Kamath H (2015) Dynamic control and optimization of distributed energy resources in a microgrid. IEEE Trans Smart Grid 6(6):2884–2894
4. Hans CA, Braun P, Raisch J, Grüne L, Reincke-Collon C (2019) Hierarchical distributed model predictive control of interconnected microgrids. IEEE Trans Sustain Energy 10(1):407–416
5. Sampathirao AK, Hans CA, Raisch J (2008) Two-stage consensus-based distributed MPC for interconnected microgrids, pp 1–8. arXiv:1810.03361
6. Kar S, Hug G, Mohammadi J, Moura JMF (2014) Distributed state estimation and energy management in smart grids: a consensus+innovations approach. IEEE J Sel Topics Signal Process 8(6):1022–1038
7. Hug G, Kar S, Wu C (2015) Consensus + innovations approach for distributed multiagent coordination in a microgrid. IEEE Trans Smart Grid 6(4):1893–1903
8. Baker K, Guo J, Hug G, Li X (2016) Distributed MPC for efficient coordination of storage and renewable energy sources across control areas. IEEE Trans Smart Grid 7(2):992–1001
9. Kargarian A, Mohammadi J, Guo J, Chakrabarti S, Barati M, Hug G, Kar S, Baldick R (2018) Toward distributed/decentralized DC optimal power flow implementation in future electric power systems. IEEE Trans Smart Grid 9(4):2574–2594
10. Nedić A, Ozdaglar A, Parrilo PA (2010) Constrained consensus and optimization in multi-agent networks. IEEE Trans Autom Control 55(4):922–938
11. Nedić A, Olshevsky A (2015) Distributed optimization over time-varying directed graphs. IEEE Trans Autom Control 60(3):601–615
12. Nedić A, Olshevsky A, Shi W (2017) Achieving geometric convergence for distributed optimization over time-varying graphs. SIAM J Optim 27(4):2597–2633
13. Nedić A, Liu J (2018) Distributed optimization for control. Annu Rev Control Robot Auton Syst 1(1):77–103
14. Tsai S, Tseng Y, Chang T (2017) Communication-efficient distributed demand response: a randomized ADMM approach. IEEE Trans Smart Grid 8(3):1085–1095
15. Scutari G, Sun Y (2018) Distributed nonconvex constrained optimization over time-varying digraphs, pp 1–44. arXiv:1809.01106v1
16. Gu C, Wu Z, Li J, Guo Y (2018) Distributed convex optimization with coupling constraints over time-varying directed graphs, pp 1–25. arXiv:1805.07916
17. Aybat NS, Hamedani EY (2019) A distributed ADMM-like method for resource sharing over time-varying networks, pp 1–28. arXiv:1611.07393v3
18. Wei E, Ozdaglar A (2013) On the O(1/k) convergence of asynchronous distributed alternating direction method of multipliers, pp 1–30. arXiv:1307.8254
19. Chang T, Hong M, Liao W, Wang X (2016) Asynchronous distributed ADMM for large-scale optimization—Part I: Algorithm and convergence analysis. IEEE Trans Signal Process 64(12):3118–3130
20. Hong M, Chang T (2017) Stochastic proximal gradient consensus over random networks. IEEE Trans Signal Process 65(11):2933–2948
21. Bastianello N, Carli R, Schenato L, Todescato M (2019) Asynchronous distributed optimization over lossy networks via relaxed ADMM: stability and linear convergence, pp 1–16. arXiv:1901.09252
22. Bastianello N, Todescato M, Carli R, Schenato L (2018) Distributed optimization over lossy networks via relaxed peaceman-rachford splitting: a robust ADMM approach. In: Proceedings of the European control conference (ECC), Limassol, Cyprus, 2018, pp 477–482
23. Majzoobi L, Shah-Mansouri V, Lahouti F (2018) Analysis of distributed ADMM algorithm for consensus optimisation over lossy networks. IET Signal Process 12(6):786–794
24. Abbe E (2017) Community detection and stochastic block models: recent developments. J Mach Learn Res 18(1):6446–6531

Chapter 6
Implementation of Distributed Methods for the MPC-Based Dispatch Scheme

This chapter is devoted to showing the performance of the distributed algorithms proposed in Chap. 5 as they are implemented in the MPC-based economic dispatch scheme of large-scale energy systems. Since the MPC-based scheme is an online optimization-based approach, the optimization algorithm used must be able to find a satisfying solution in a given period of time. In this regard, additional techniques to improve the convergence speed of the algorithm can be added. Particularly, Nesterov's acceleration and the warm start methods are considered. Therefore, they are discussed in Sects. 6.1 and 6.2, respectively. Afterward, a numerical simulation study is carried out using the benchmark case and presented in Sect. 6.3. Finally, Sect. 6.4 concludes this chapter with some remarks.

6.1 Accelerated Versions

This section discusses the application of Nesterov's accelerated gradient method to Algorithms 5.1 and 5.2. First, Nesterov's acceleration method and the accelerated version of Algorithms 5.1 and 5.2 are presented. Then, some numerical results, which show the improvement on the convergence speed obtained by implementing the acceleration method, are provided.

6.1.1 Adapting Nesterov's Accelerated Gradient Method

Nesterov proposes an optimal gradient method that has convergence rate of $\mathcal{O}(\frac{1}{k^2})$ for a convex smooth optimization in [1]. The main idea of this method is that the gradient step is taken from a smartly chosen interpolated point of the last two iterations. This method has been further extended for non-smooth cases in [2]. Furthermore, a unifying framework to analyze a generalization of this method and its variants is

W. W. Ananduta, *Non-centralized Optimization-Based Control Schemes for Large-Scale Energy Systems*, Springer Theses,
https://doi.org/10.1007/978-3-030-89803-8_6

provided in [3]. As an MPC technique, this approach is discussed in [4, 5], where the primal cost function is assumed to be strongly convex and the dual problem is solved. Note that, [4] considers a centralized scheme, whereas [5] discusses a distributed one. Similarly, in [6], network resource allocation problems with strongly concave cost function is solved in a distributed manner using an accelerated dual ascent method, i.e., the update of the dual variables uses the Nesterov's accelerated gradient method. Additionally, an ADMM algorithm that implements this technique for a more general convex problems with an equality constraint is presented in [7, 8].

To show the accelerated gradient method, consider the following problem:

$$\underset{u}{\text{minimize}}\ f(u),$$

where $u \in \mathbb{R}^{n_u}$ and $f(u) : \mathbb{R}^{n_u} \to \mathbb{R}$ is a differentiable convex function wih Lipschitz continuous gradient. Moreover, denote by ω the Lipschitz constant of the gradient of $f(u)$, which is denoted by $\nabla f(u)$. The accelerated gradient method for the preceding problem, as presented in [8], is shown in Algorithm 6.1. As can be seen, $\hat{u}(k)$ is the interpolated point that is used to perform the gradient step.

Algorithm 6.1 Nesterov's accelerated gradient descent

Initialization: $\theta(0) = 1$, $u(0) = u(-1) \in \mathbb{R}^{n_u}$, $\eta \leq 1/\omega$.

Iteration:

1. Update $\theta(k+1)$ by $\theta(k+1) = \frac{1}{2}(1 + \sqrt{4\theta(k)^2 + 1})$
2. Update $\hat{u}(k+1)$ by $\hat{u}(k+1) = u(k) + \frac{\theta(k)-1}{\theta(k+1)}(u(k) - u(k-1))$
3. Update $u(k+1)$ by $u(k+1) = \hat{u}(k+1) - \eta \nabla f(\hat{u}(k+1))$

Now, the accelerated technique is adapted to the stochastic DAL and ADMM-based methods as shown in Algorithms 6.2 and 6.3 (pages 85 and 86, respectively). Since in these algorithms the gradient step is applied to the dual variable updates, an interpolated point of the dual variable must be computed at each iteration based on the rule shown in steps 1 and 2 of Algorithm 6.1. Due to this requirement, all agents must always be active at each iteration, i.e., the following assumption holds.

Assumption 6.1 The probability of agent $p \in \mathcal{P}$ being active is 1, i.e., $\mathbb{P}(p \in \mathcal{A}(k)) = 1$, for all $p \in \mathcal{P}$ and $k \in \mathbb{Z}_{\geq 0}$. □

The preceding assumption implies that Algorithms 6.2 and 6.3 are performed synchronously. Nevertheless, the communication network might still be randomly time-varying. Notice that in Algorithms 6.2 and 6.3, the interpolated point $\hat{\boldsymbol{\lambda}}_i(k+1)$ and $\hat{\boldsymbol{\mu}}_i(k+1)$, respectively, are used to update the primal variables (see step 3 of both algorithms). Additionally, in Algorithm 6.2, due to the initialization of the dual variables, the auxiliary variables to keep tracking the dual variables of the neighbors are not needed, since $\boldsymbol{\lambda}_i^j(k) = \boldsymbol{\lambda}_j^i(k)$, for all $\{i, j\} \in \mathcal{E}$ and $k \in \mathbb{Z}_{\geq 0}$ (c.f. Remark 5.1).

It is also noted that the acceleration technique is treated as a heuristic method and the effectiveness will be shown through numerical simulations. Although [7, 8] show

Algorithm 6.2 Stochastic accelerated DAL method

Initialization: For each node $i \in \mathcal{N}$, $\boldsymbol{v}_i(0) = \boldsymbol{v}_{i0} \in \mathbb{R}^{|\mathcal{N}_i|h}$ and $\boldsymbol{\lambda}_i(0) = \boldsymbol{\lambda}_i(-1) = 0 \in \mathbb{R}^{|\mathcal{N}_i|h}$. Moreover, $z_i^j(0) = \boldsymbol{v}_j^i(0)$ and $\boldsymbol{\xi}_i^j(0) = \boldsymbol{\lambda}_j^i(0)$, for all $j \in \mathcal{N}_i$ and $i \in \mathcal{N}$, and $\theta(0) = 1$.
Iteration: For each agent $p \in \mathcal{P}$,

1. Update $\theta(k+1)$ as follows:
$$\theta(k+1) = \frac{1}{2}(1 + \sqrt{4\theta(k)^2 + 1}).$$
2. Update $\hat{\boldsymbol{\lambda}}_i(k+1)$, for all $i \in \mathcal{M}_p$, as follows:
$$\hat{\boldsymbol{\lambda}}_i(k+1) = \boldsymbol{\lambda}_i(k) + \frac{\theta(k) - 1}{\theta(k+1)}(\boldsymbol{\lambda}_i(k) - \boldsymbol{\lambda}_i(k-1)).$$
3. Update $\left(\boldsymbol{u}_i(k+1), \hat{\boldsymbol{v}}_i(k)\right)$, for all $i \in \mathcal{M}_p$, according to
$$\{\left(\boldsymbol{u}_i(k+1), \hat{\boldsymbol{v}}_i(k)\right)\}_{i\in\mathcal{M}_p} = \arg \min_{\{(\boldsymbol{u}_i,\boldsymbol{v}_i)\in\mathcal{L}_i\}_{i\in\mathcal{M}_p}} \sum_{i\in\mathcal{M}_p} \Bigg(f_i^{\mathrm{l}}(\boldsymbol{u}_i) + f_i^{\mathrm{c}}(\boldsymbol{v}_i) + \\ + \sum_{j\in\mathcal{N}_i} \left(2\langle \hat{\boldsymbol{\lambda}}_i^j(k+1), \boldsymbol{v}_i^j \rangle + \|\boldsymbol{v}_i^j + z_i^j(k)\|_2^2 \right) \Bigg).$$
4. Update $\boldsymbol{v}_i^j(k+1)$, for all $j \in \mathcal{N}_i$ and $i \in \mathcal{M}_p$, as follows:
$$\boldsymbol{v}_i^j(k+1) = \begin{cases} \eta_i^j \hat{\boldsymbol{v}}_i^j(k) + \left(1 - \eta_i^j\right) \boldsymbol{v}_i^j(k), & \text{if } \phi(j,\cdot) \in \mathcal{A}_p(k+1) \cup \{p\}, \\ \boldsymbol{v}_i^j(k), & \text{otherwise.} \end{cases}$$
5. For each $j \in \mathcal{N}_i \backslash \mathcal{M}_p$ such that $\phi(j,\cdot) \in \mathcal{A}_p(k+1)$ and $i \in \mathcal{M}_p$, send $\boldsymbol{v}_i^j(k+1)$ to and receive $\boldsymbol{v}_j^i(k+1)$ from agent $\phi(j,\cdot)$.
6. Update the auxiliary and dual variables $z_i(k+1)$ and $\boldsymbol{\lambda}_i(k+1)$, for all $i \in \mathcal{M}_p$, according to
$$z_i^j(k+1) = \begin{cases} \boldsymbol{v}_j^i(k+1), & \text{if } \phi(j,\cdot) \in \mathcal{A}_p(k+1) \cup \{p\}, \\ z_i^j(k), & \text{otherwise,} \end{cases}$$
$$\boldsymbol{\lambda}_i^j(k+1) = \begin{cases} \hat{\boldsymbol{\lambda}}_i^j(k+1) + \eta_i^j\left(\boldsymbol{v}_i^j(k+1) + z_i^j(k+1)\right), & \text{if } \phi(j,\cdot) \in \mathcal{A}_p(k+1) \cup \{p\}, \\ \hat{\boldsymbol{\lambda}}_i^j(k+1), & \text{otherwise.} \end{cases}$$

---

the convergence rate analysis of an accelerated ADMM method, the problems considered in those papers are limited and do not include the economic dispatch problem (3.1). Specifically, [8], which provides a more general result than [7], shows the convergence rate of $\mathcal{O}(\frac{1}{k^2})$ when the cost function is composed of two component, one is strongly convex while the other is quadratic. Moreover, the problems considered only have an equality constraint but do not include local constraint sets.

Algorithm 6.3 Stochastic accelerated distributed ADMM-based method

Initialization: For each $i \in \mathcal{N}$, $\boldsymbol{\mu}_i(0) = \boldsymbol{\mu}_{i0} \in \mathbb{R}^{|\mathcal{N}_i|h}$, and $\boldsymbol{y}_i^j(0) = \boldsymbol{y}_j^i(0) \in \mathbb{R}^h$, for all $j \in \mathcal{N}_i$, and $\theta(0) = 1$.
Iteration: For each agent $p \in \mathcal{P}$,

1. Update $\theta(k+1)$ as follows:
$$\theta(k+1) = \frac{1}{2}(1 + \sqrt{4\theta(k)^2 + 1}).$$
2. Update $\hat{\boldsymbol{\mu}}_i(k+1)$, for all $i \in \mathcal{M}_p$, as follows:
$$\hat{\boldsymbol{\mu}}_i(k+1) = \boldsymbol{\lambda}_i(k) + \frac{\theta(k) - 1}{\theta(k+1)}(\boldsymbol{\mu}_i(k) - \boldsymbol{\mu}_i(k-1)).$$
3. Update $\boldsymbol{u}_i(k+1)$ and $\boldsymbol{v}_i(k+1)$, for all $i \in \mathcal{M}_p$, according to
$$\{(\boldsymbol{u}_i(k+1), \boldsymbol{v}_i(k+1))\}_{i\in\mathcal{M}_p} = \arg \min_{\{(\boldsymbol{u}_i,\boldsymbol{v}_i)\in\mathcal{L}_i\}_{i\in\mathcal{M}_p}} \sum_{i\in\mathcal{M}_p} \Big(f_i^{\mathrm{l}}(\boldsymbol{u}_i) + f_i^{\mathrm{c}}(\boldsymbol{v}_i) + \sum_{j\in\mathcal{N}_i} \langle \hat{\boldsymbol{\mu}}_i^j(k+1), \boldsymbol{v}_i^j \rangle + \frac{1}{2}\|\boldsymbol{v}_i^j - \boldsymbol{y}_i^j(k)\|_2^2 \Big).$$
4. For each $j \in \mathcal{N}_i \backslash \mathcal{M}_p$ such that $\phi(j,\cdot) \in \mathcal{A}_p(k+1)$ and $i \in \mathcal{M}_p$, send $\boldsymbol{v}_i^j(k+1)$ and $\boldsymbol{\mu}_i^j(k)$ to agent $\phi(j,\cdot) \in \mathcal{N}_p^{\mathrm{s}}$ and receive $\boldsymbol{v}_j^i(k+1)$ and $\boldsymbol{\mu}_j^i(k)$ from agent $\phi(j,\cdot) \in \mathcal{N}_p^{\mathrm{s}}$.
5. Update $\boldsymbol{y}_i^j(k+1)$, for all $j \in \mathcal{N}_i$ and $i \in \mathcal{M}_p$, as follows:
$$\boldsymbol{y}_i^j(k+1) = \begin{cases} \frac{1}{2}\Big(\boldsymbol{\mu}_i^j(k) - \boldsymbol{\mu}_j^i(k) + \boldsymbol{v}_i^j(k+1) - \boldsymbol{v}_j^i(k+1)\Big), \\ \qquad \text{if } \phi(j,\cdot) \in \mathcal{A}_p(k+1) \cup \{p\}, \\ \boldsymbol{y}_i^j(k), \quad \text{otherwise.} \end{cases}$$
6. Update the dual variables $\boldsymbol{\mu}_i^j(k+1)$, for all $j \in \mathcal{N}_i$ and $i \in \mathcal{M}_p$, according to
$$\boldsymbol{\mu}_i^j(k+1) = \begin{cases} \hat{\boldsymbol{\mu}}_i^j(k{+}1) {+} \eta_i^j\Big(\boldsymbol{v}_i^j(k{+}1) {-} \boldsymbol{y}_i^j(k{+}1)\Big), & \text{if } \phi(j,\cdot){\in}\mathcal{A}_p(k{+}1){\cup}\{p\}, \\ \hat{\boldsymbol{\mu}}_i^j(k{+}1), & \text{otherwise.} \end{cases}$$

6.1.2 *Convergence Speed Improvement*

The improvement obtained by applying Nesterov's accelerated gradient method is shown through numerical simulations. Therefore, consider the network of prosumers presented in Section 5.5. However, in this simulation study, Algorithms 5.1, 5.2, 6.2, and 6.3 are applied to a number of randomly generated cases, in each of which, the network $\mathcal{G}$ is randomly generated based on the Erdös-Rényi model with the number of nodes and the connectivity level (the proportion of the number of links with respect to the maximum possible number of links) as the parameters that are varied. Furthermore, at each scenario, the upper-bound of u_i, the load d_i, and the cost

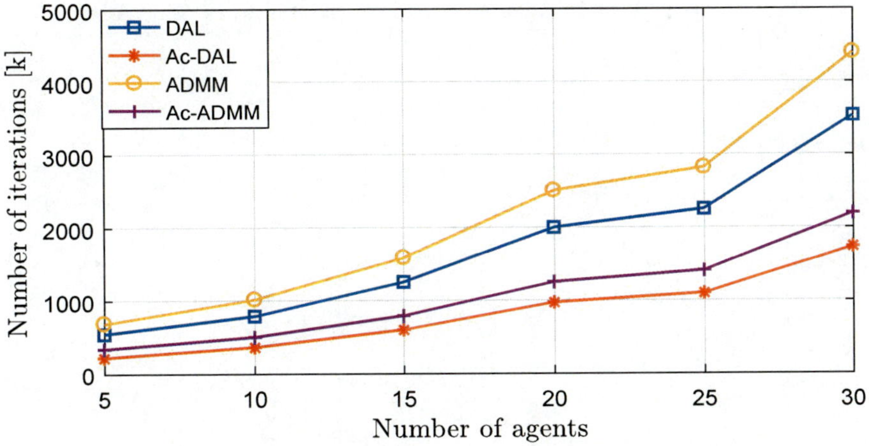

Fig. 6.1 Performance of the proposed algorithms on networks with different number of nodes

coefficient c_i^{dg}, for each node $i \in \mathcal{N}$, are sampled with uniform distribution, i.e., $u_i^{\max} \sim 10 + [-1, 1]$, $d_i \sim [8.25] + [-0.25, 0.25]$, and $c_i^{\mathrm{dg}} \sim [7.5] + [-2.5, 2.5]$. Moreover, at each case, it is set that $h = 1$, $\beta_{ij} = 0.9$, for all $j \in \mathcal{N}_i$ and $i \in \mathcal{N}$, whereas $\mathcal{A}(k) = \mathcal{P}$, for all $k \in \mathbb{Z}_{\geq 0}$. Additionally, the DAL algorithms 5.1 and 6.2 use $\eta_{ij} = 0.2$, for all $j \in \mathcal{N}_i$ and $i \in \mathcal{N}$, whereas, the ADMM-based algorithms 5.2 and 6.3 use $\eta_{ij} = 0.6$, for all $j \in \mathcal{N}_i$ and $i \in \mathcal{N}$.

Figure 6.1 shows how these four algorithms performed on the networks with 5, 10, 15, 20, 25, and 30 nodes and 40% of connectivity level. Moreover, Fig. 6.2 shows the simulation results where the connectivity level is varied and the number of nodes is fixed. As can be seen, the accelerated versions, Algorithms 6.2 and 6.3, which are denoted by Ac-DAL and Ac-ADMM, respectively, require less number of iterations to stop, i.e., to reach the tolerable level of primal residual error, than the standard versions (Algorithms 5.1 and 5.2). Additionally, Fig. 6.1 also shows how the number of nodes in the network influence the number of iterations required to reach the same performance level. As expected, the number of iterations required grows with the number of nodes due to the increasing dimension of decision variables. However, based on Fig. 6.2, the number of iterations does not have a positive correlation with the connectivity level as that with the number of nodes.

6.2 Warm Start Method

Another technique to reduce the number of iterations performed by the distributed algorithms is the warm start method [9]. As suggested by its name, the idea of this method is carefully choosing the initial condition of the variables. Particularly for an MPC scheme, where a similar problem is repeatedly solved at each time step, the

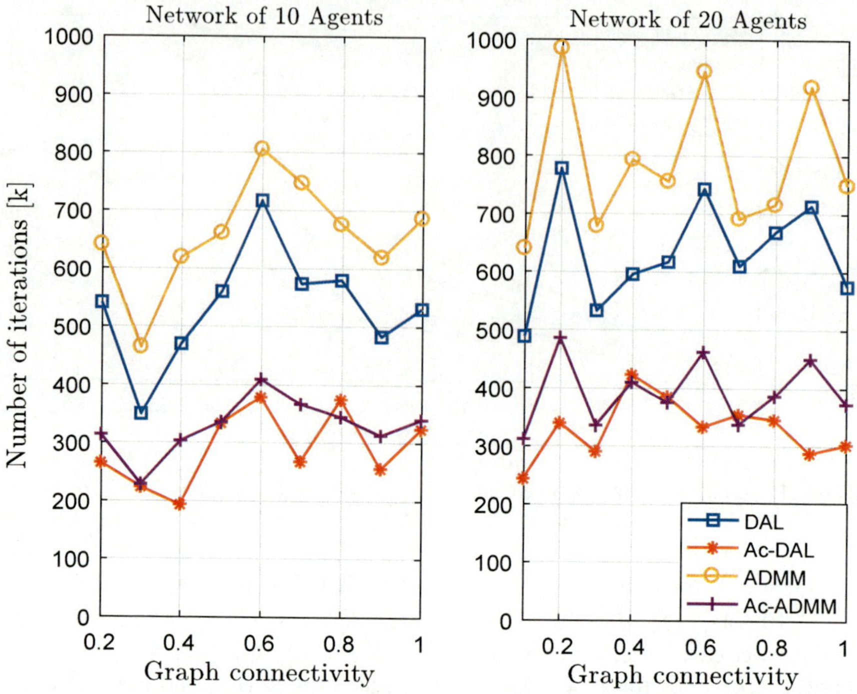

Fig. 6.2 Performance of the proposed algorithms on networks with different level of connectivity. Left plot considers a 10-node network whereas right plot considers a 20-node network

variables can be initialized using the computed value from the previous time step. Note that the difference between the problem solved at one step and another is the uncertainty that comes from the load and non-dispatchable generation. At time step t, although only $u_{i,t}$ and $v_{i,t}$ are implemented, the decisions for h time steps ahead, i.e., $u_{i,\tau}$ and $v_{i,\tau}$,for all $\tau \in \mathcal{T}$, are computed. Furthermore, the dual variables from the last iteration at t can also be saved to be also used at the next time step, $t+1$.

Now, consider the DAL algorithm and let $\boldsymbol{v}^j_{i,t|t} = \text{col}(\{v^j_{i,\tau|t}\}_{\tau\in\mathcal{T}})$ and $\boldsymbol{\lambda}^j_{i,t|t} = \text{col}(\{\lambda^j_{i,\tau|t}\}_{\tau\in\mathcal{T}})$, for all $j \in \mathcal{N}_i$ and $i \in \mathcal{N}$, denote the coupled and dual solutions computed at time t. Therefore, in the next time step, $t+1$, they can be initialized as follows:

$$\boldsymbol{v}^j_{i,t+1}(0) = \text{col}(\{\boldsymbol{v}^j_{i,\tau|t}\}^{t+h-1}_{\tau=t+1}, \boldsymbol{v}^j_{i,t+h-1|t}),$$
$$\boldsymbol{\lambda}^j_{i,t+1}(0) = \text{col}(\{\boldsymbol{\lambda}^j_{i,\tau|t}\}^{t+h-1}_{\tau=t+1}, \boldsymbol{\lambda}^j_{i,t+h-1|t}),$$

for all $j \in \mathcal{N}_i$ and $i \in \mathcal{N}$. Similarly, let the auxiliary and dual variables at the end of the iteration of the ADMM-based algorithm at time step t be denoted by $\boldsymbol{y}^j_{i,t|t} = \text{col}(\{y^j_{i,\tau|t}\}_{\tau\in\mathcal{T}})$ and $\boldsymbol{\mu}^j_{i,t|t} = \text{col}(\{\mu^j_{i,\tau|t}\}_{\tau\in\mathcal{T}})$, for all $j \in \mathcal{N}_i$ and $i \in \mathcal{N}$. Then, in the next time step $t+1$, these variables are initialized as follows:

$$\begin{aligned}
\boldsymbol{y}_{i,t+1}^{j}(0) &= \mathrm{col}(\{\boldsymbol{y}_{i,\tau|t}^{j}\}_{\tau=t+1}^{t+h-1}, \boldsymbol{y}_{i,t+h-1|t}^{j}), \\
\boldsymbol{\mu}_{i,t+1}^{j}(0) &= \mathrm{col}(\{\boldsymbol{\mu}_{i,\tau|t}^{j}\}_{\tau=t+1}^{t+h-1}, \boldsymbol{\mu}_{i,t+h-1|t}^{j}),
\end{aligned}$$

for all $j \in \mathcal{N}_i$ and $i \in \mathcal{N}$. Notice that the variables associated with the last time step in the horizon, i.e., $t + h$, are initialized in the same way as the variables at $t + h - 1$. The reason of this choice is twofold. Firstly, as previously explained, the difference between the optimization problem solved at t and $t + 1$ is the uncertain variables d_i, for each $i \in \mathcal{N}$. Secondly, the difference between $d_{i,t+h}$ and $d_{i,t+h-1}$, for each $i \in \mathcal{N}$, might not be large, especially when the sampling period is small.

6.3 Simulation Study of DMPC Scheme

This section presents the simulation study, where Algorithms 5.1 and 5.2 as well as their accelerated versions (Algorithms 6.2 and 6.3, respectively) are implemented in the DMPC scheme. First, the setup of the test case is explained and then the simulation results are presented and discussed.

6.3.1 Simulation Setups

The benchmark case, which is described in Sect. 2.4, is considered in this simulation study. The locations of dispatchable generators, non-dispatchable generators, and storage units in the network are shown in Fig. 6.3. The parameters of the components in the network are given in Table 6.2. Moreover, the available load data set is used as the maximum value of the load at each node. The nodes that have a maximum load greater than 100 kW are considered to have a commercial load profile. Otherwise,

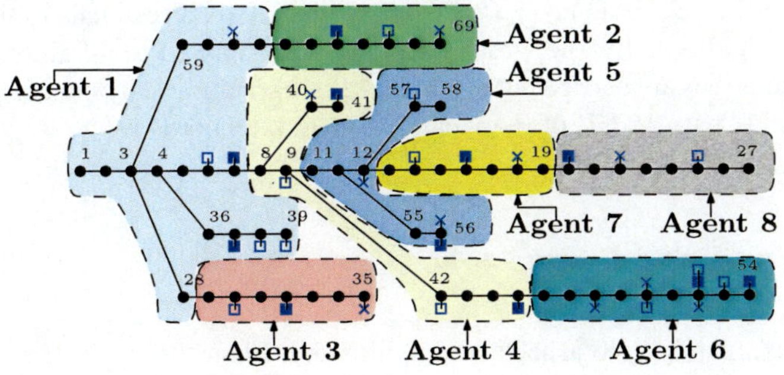

Fig. 6.3 The locations of non-dispatchable units (indicated by blue filled squares), dispatchable generators (indicated by blue empty squares), and storage units (indicated by blue crosses) in the benchmark case

Table 6.1 Maximum generated power of solar-based generation units

Nodes	Maximum power (kW)
7, 32, 37, 45, 56, 65	50
20, 41	75
16, 54	100
52	150

they have a residential load profile. The non-dispatchable generation units are solar-based, with the maximum power generation as shown in Table 6.1. Additionally, it is assumed that the subsystems have the knowledge of the loads and non-dispatchable power generation, implying perfect forecast so that the analysis can be focused only on the outcomes of the algorithms. An example of residential and commercial load profiles is shown in the top plot of Fig. 6.4, whereas a solar-based generation profile is shown in the bottom plot of Fig. 6.4 (Table 6.2).

The simulation time is one day with the sampling time of 15 min, implying 96 steps. Moreover, the prediction horizon in the DMPC scheme is $h = 4$. Furthermore, it is assumed that the probability of each communication link being active is equal, i.e., $\beta_{pq} = \beta = 0.9$, for each $\{p, q\} \in \mathcal{E}^s$, whereas all agents are always active at each iteration to accommodate the accelerated algorithms. Four simulations have been carried out and each simulation uses the same case but a different distributed algorithm, i.e., the DAL, accelerated DAL, ADMM-based, and accelerated ADMM-based methods. For the DAL and the accelerated DAL methods (Algorithms 5.1 and 6.2, respectively), the step sizes are set to be $\eta_i^j = 0.2$, for all $\{i, j\} \in \mathcal{E}$. Moreover, their stopping criterion is $\|\boldsymbol{v}_t(k) + z_t(k)\|_2 \leq 1$, which is the primal residual [9]. Note that since the stopping criterion corresponds to the coupling constraints and the maximum allowable power transferred is 200 kW, the choice of the stopping criterion is small enough and tolerable. For the ADMM-based method and its accelerated version (Algorithms 5.2 and 6.3, respectively), the step sizes are set to be $\eta_i^j = 0.6$, for all $\{i, j\} \in \mathcal{E}$. Additionally, the stopping criterion used in these algorithms is also the primal residual, i.e., $\|\boldsymbol{v}_t(k) - \boldsymbol{y}_t(k)\|_2 \leq 1$, which also corresponds to the coupling constraints. Additionally, the warm start method is applied to all algorithms. Note that all simulations are carried out in MATLAB with Yalmip [10] and the `quadprog` solver on a PC with 16 GB of RAM and 2.6 GHz Intel core i7.

6.3.2 *Simulation Results*

In this study, mainly, two aspects are highlighted. First, the improvement obtained from implementing the acceleration technique. Second, a comparison of the performance of the DAL and the ADMM methods. Figures 6.5, 6.6, 6.7 and 6.8 show the simulation results.

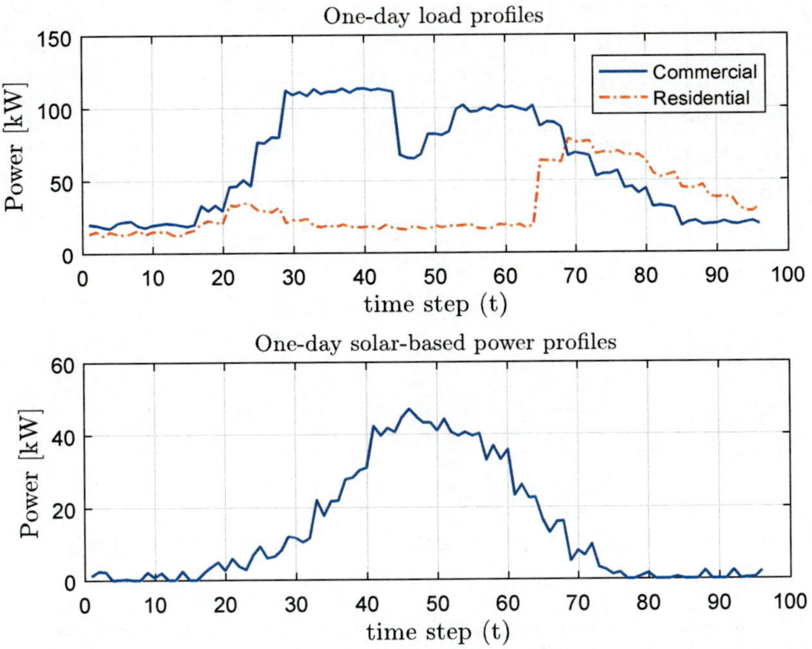

Fig. 6.4 One-day residential and commercial load profiles (top plot) and one-day solar-based power generation profile (bottom plot)

Table 6.2 Parameters of the network components

Parameters	Value	Unit	Bus
$u_i^{\mathrm{dg,min}}, u_i^{\mathrm{dg,max}}$	0, 400	kW	$i \in \mathcal{N}^{\mathrm{dg}}$
$x_i^{\mathrm{min}}, x_i^{\mathrm{max}}, x_{i,0}$	30, 100, 50	%	$i \in \mathcal{N}^{\mathrm{st}}$
$u_i^{\mathrm{ch}}, u_i^{\mathrm{dh}}$	100, 100	kW	$i \in \mathcal{N}^{\mathrm{st}}$
$e_{\mathrm{cap},i}$	1000	kWh	$i \in \mathcal{N}^{\mathrm{st}}$
a_i	1	–	$i \in \mathcal{N}^{\mathrm{st}}$
$v_i^{j,\mathrm{max}}$	200	–	$\{i, j\} \in \mathcal{E}$
$c_i^{\mathrm{st}}, c_i^{\mathrm{dg}}$	1, 10	–	$i \in \mathcal{N}$
$c_i^{\mathrm{tp}}, c_i^{\mathrm{t}}$	10, 1	–	$i \in \mathcal{N}$

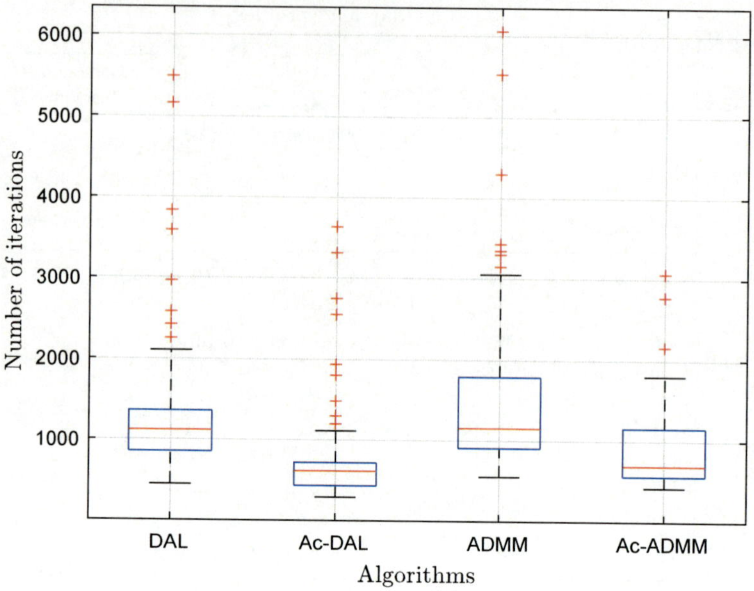

Fig. 6.5 The number of iterations performed by each algorithms. The blue boxes show the 25th percentiles until the 75th percentiles and the red lines show the median

Figure 6.5 shows the number of iterations required by the stochastic DAL, accelerated DAL, ADMM-based, and accelerated ADMM-based methods, over all time steps. While the red lines show the median, the blue boxes cover the 25th–75th percentiles. The average number of iterations are 1294.9, 730.6, 1518.8, and 905.4, respectively. From these results and together with the results presented in Sect. 6.1.2, it can clearly be concluded that adding the acceleration technique to both DAL and ADMM-based, algorithms improve the convergence speed. However, it is also important to note that the acceleration technique does not allow the subsystems to be inactive randomly since they must always perform the interpolation steps at each iteration.

In order to compare the performance of the DAL and ADMM-based methods, first, the power imbalances are evaluated. Based on (2.1) and (2.2), the coupled and local power imbalance indices, denoted by Δ_t^{l} and Δ_t^{c}, respectively, are defined as follows:

$$\Delta_t^{\mathrm{c}} = \mathrm{col}(\{v_{i,t}^j + v_{j,t}^i\}_{j\in\mathcal{N}_i, i\in\mathcal{N}}), \tag{6.1}$$

$$\Delta_t^{\mathrm{l}} = \mathrm{col}(\{\mathbb{1}_{n_i^{\mathrm{l}}}^\top u_{i,t} + \mathbb{1}_{|\mathcal{N}_i|}^\top v_{i,t} - d_{i,t}\}_{i\in\mathcal{N}}). \tag{6.2}$$

Fig. 6.6 shows $\|\Delta_t^{\mathrm{c}}\|_2$ (top plot) and $\|\Delta_t^{\mathrm{l}}\|_2$ (bottom plot). It can be observed that $\|\Delta_t^{\mathrm{c}}\|_2$ obtained from the DAL methods is lower than that of the ADMM-based methods, most of the time. The lower $\|\Delta_t^{\mathrm{c}}\|_2$ is, the better since it implies the satis-

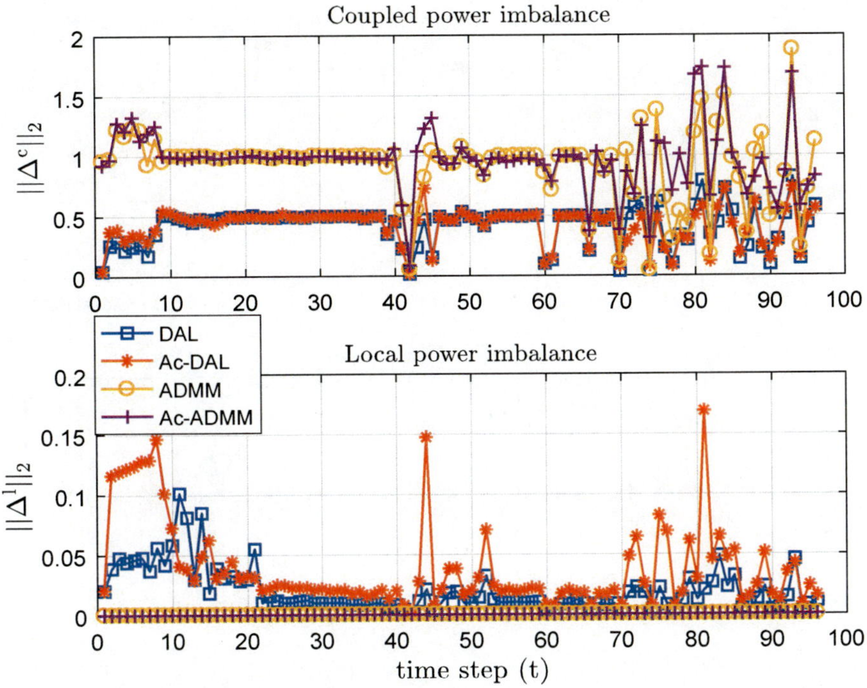

Fig. 6.6 Top plot shows the coupled power imbalance, $||\Delta^c||_2$ (see (6.1)). Bottom plot shows the local power imbalance, $||\Delta^l||_2$ (see (6.2))

faction of the coupling constraints. Therefore, to achieve the same performance, the stopping criterion of the ADMM-based method must be set lower, implying a larger number of iterations than the current simulation results. Moreover, in terms of local power imbalance, $\|\Delta_t^l\|_2$ obtained from the ADMM-based methods is zero since both $u_{i,t}$ and $v_{i,t}$, for all $i \in \mathcal{N}$ and $t \in \mathbb{Z}_{\geq 0}$, are computed by solving the local optimizations, which include the local power balance constraints (2.2). On the other hand, in the DAL methods, $v_{i,t}$, for all $i \in \mathcal{N}$ and $t \in \mathbb{Z}_{\geq 0}$, are obtained by performing a convex combination step of the solution of the local optimization and the value at the previous iteration. This explains the non-zero value of the local power imbalance Δ_t^l.

Figure 6.7 shows the total transferred power among the subsystems. As can be seen from the top plot, the subsystems in the network are actively transferring power among each other in order to perform optimally. Furthermore, Fig. 6.8 shows the cost at each iteration, for all methods, which differ due to the early termination of the algorithms.

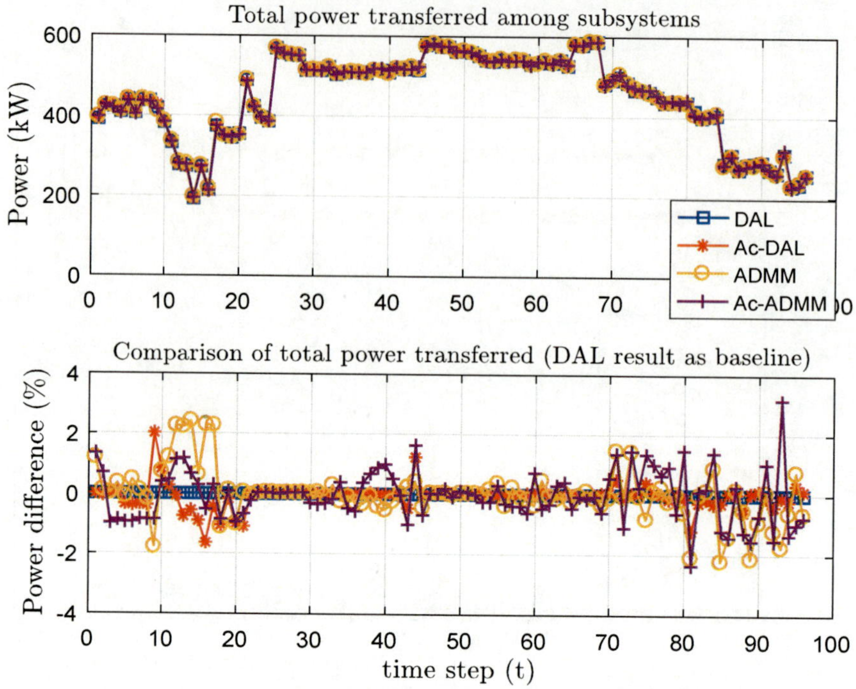

Fig. 6.7 The total power transferred among subsystems at each iteration. Top plot shows the total power transferred while bottom plot compares the difference of the transferred power obtained by different algorithms with the result of the DAL method as the baseline

6.4 Summary

This chapter presents a simulation study of the implementation of the DAL and ADMM-based distributed methods to the MPC-based economic dispatch scheme. Moreover, in order to improve the convergence speed of these methods, Nesterov's acceleration technique is added to the algorithms and the simulation results show the effectiveness of this acceleration technique. By implementing the proposed distributed algorithms, the non-centralized MPC scheme is robust against random communication failures. This feature is important considering its application to energy systems. Finally, note that in this chapter, both the DAL and ADMM-based algorithms are implemented in a fully distributed MPC scheme, which requires intensive neighbor-to-neighbor communication. Next, in Chap. 7, a different non-centralized scheme, which requires less intensive communication, is proposed.

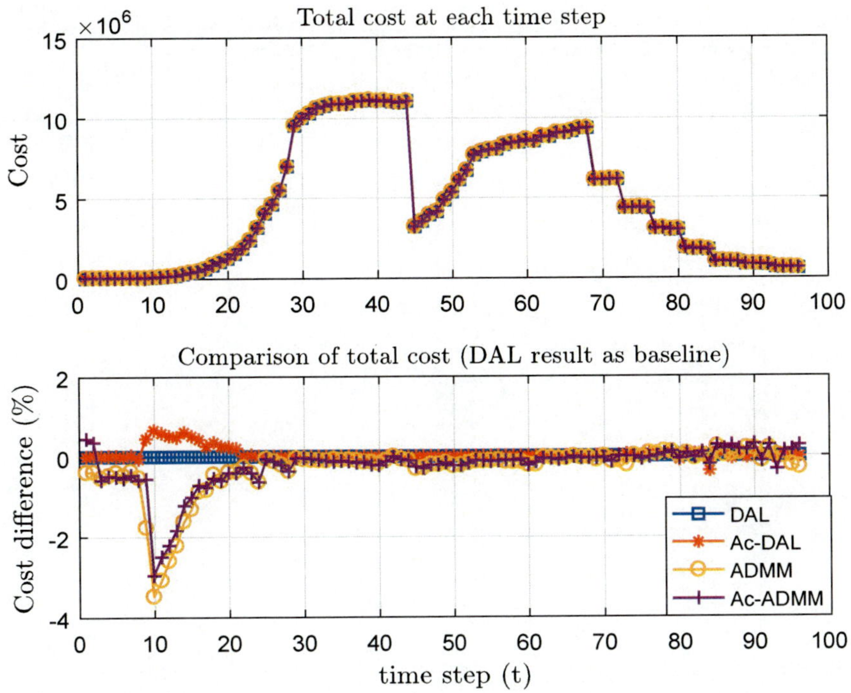

Fig. 6.8 The stage cost at each iteration. Top plot shows the cost while bottom plot compares the difference of the cost obtained by different algorithms with the result of the DAL method as the baseline

References

1. Nesterov Y (1983) A method for solving the convex programming problem with convergence rate $O(1/k^2)$. Dokl. Akad. Nauk SSSR, 27:543–547 (translated as Sov. Math, Dokl)
2. Beck A, Teboulle M (2009) A fast iterative shrinkage-thresholding algorithm for linear inverse problems. SIAM J Imaging Sci 2(1):183–202
3. Tseng P (submitted) On accelerated proximal gradient methods for convex-concave optimization. SIAM J Optim
4. Patrinos P, Bemporad A (2013) An accelerated dual gradient-projection algorithm for embedded linear model predictive control. IEEE Trans Autom Control 59(1):18–33
5. Giselsson P, Doan MD, Keviczky T, De Schutter B, Rantzer A (2013) Accelerated gradient methods and dual decomposition in distributed model predictive control. Automatica 49(3):829–833
6. Beck A, Nedić A, Ozdaglar A, Teboulle M (2014) An $o(1/k)$ gradient method for network resource allocation problems. IEEE Trans Control of Network Syst 1(1):64–73
7. Goldfarb D, Ma S, Scheinberg K (2013) Fast alternating linearization methods for minimizing the sum of two convex functions. Math Program 141(1–2):349–382
8. Goldstein T, O'Donoghue B, Setzer S, Baraniuk R (2014) Fast alternating direction optimization methods. SIAM J Imag Sci 7(3):1588–1623
9. Boyd S, Parikh N, Chu E, Peleato B, Eckstein J (2011) Distributed optimization and statistical learning via the alternating direction method of multipliers. Found Trends in Mach Learn 3(1):1–122

10. Löfberg J (2004) YALMIP: a toolbox for modeling and optimization in MATLAB. In: Proceedings of the CACSD conference, Taipei, Taiwan, pp 284–289

Chapter 7
A Non-centralized Control Scheme Based on Online Partitioning

In this chapter, a novel non-centralized economic dispatch scheme is proposed. It is inspired by the coalitional control scheme and is a mixture of distributed and decentralized ones. The methodology, which is outlined in Sect. 7.2, combines an online repartitioning method (Sect. 7.3) and a procedure to form self-sufficient coalitions of subsystems to solve the economic dispatch problem (2.3) (Sect. 7.4). Furthermore, the analysis of the methodology, including the outcomes of the repartitioning and coalition formation algorithms, is presented in the corresponding sections. Moreover, an upper bound of the suboptimality of the proposed scheme is also provided in Sect. 7.5. Some numerical simulations are also carried out to show the effectiveness of the proposed scheme in the benchmark case, as presented in Sect. 7.6. Finally, Sect. 7.7 concludes this chapter with some remarks. The methodology that is presented in this chapter is an extension of that in [1], where a periodical repartitioning scheme for a fully decentralized scheme is proposed. Additionally, a feasibility issue arisen from subsystems that are not self-sufficient, which can be found when using the scheme in [1], is resolved by the coalition-based approach proposed in this chapter. Additionally, this chapter has also been partially reported in [2].

7.1 Introduction

As discussed in Chap. 3 and [3–6], a non-centralized approach to solving MPC-based economic dispatch problems is usually based on a distributed optimization algorithm, which is iterative and require high information flow. This chapter presents an alternative non-centralized scheme with relatively low information flow and iterations, which are desirable for online optimization. The scheme consists of (i) a proper partitioning of the network and (ii) the formulation of coalition-based sub-problems that requires a coalition formation algorithm.

W. W. Ananduta, *Non-centralized Optimization-Based Control Schemes for Large-Scale Energy Systems*, Springer Theses,
https://doi.org/10.1007/978-3-030-89803-8_7

The purpose of the repartitioning scheme is to obtain self-sufficient and efficient subsystems. Loosely speaking, a subsystem is self-sufficient when it can provide its loads using its local generation units. Note that a precise definition of self-sufficiency used throughout the chapter is provided later. When this goal is achieved, each subsystem does not need to rely on the other subsystems. In addition, the efficiency criterion is in line with the objective of the economic dispatch problem.

Therefore, a distributed repartitioning procedure that has low computational burden is proposed. The proposed algorithm is a graph-partitioning-based technique and closely related to the partitioning methods presented in [1, 7]. The main idea is to move some nodes that lead to the cost improvement from one partition to another. The proposed technique belong to the class of local improvement methods, which includes the methods discussed in [8–10]. In most literature, e.g., [9, 11–15] system partitioning procedures are considered as an offline task that is carried out only once, before applying a non-centralized control approach. Differently, the approach proposed in this chapter changes the partitions of the system online. In this regard, the repartitioning of the network is triggered by a certain event, i.e., at least one subsystem is not self-sufficient.

Afterward, the economic dispatch problem of the whole system is decomposed into coalition-based sub-problems. A coalition, which consists of a number of subsystems, must be formed since the repartitioning procedure does not guarantee that the resulting subsystems are self-sufficient. In this regard, a distributed coalition formation algorithm is proposed which ensures each formed coalition is self-sufficient. Consequently, coalition-based economic dispatch sub-problems can be formulated. These problems are solved by the local controllers of the subsystems in order to obtain a (sub)-optimal solution to the centralized economic dispatch problem.

The coalition-based economic dispatch approach is inspired by the coalitional control approach [16–20]. In the coalitional control scheme, the subsystems are clustered into several coalitions based on the relevance of the subsystems, e.g., the degree of coupling among the subsystems. Furthermore, the subsystems that are in the same coalition cooperatively compute their control inputs. In the proposed method, the coalitions formed are based on the necessity to maintain the feasibility of the economic dispatch sub-problems, which can be perceived as the relevance of the agents to the problem itself. Specifically related to the control problems of energy systems, Fele et al. [19] provide a case study of voltage control in a power network. However, a coalitional scheme in the dispatch level has not been proposed in the literature.

The overall approach can be perceived as a mixture of distributed and decentralized methods. A distributed optimization algorithm is still used to solve coalition-based dispatch problems when a coalition consists of more than one subsystem, implying neighbor-to-neighbor communication is required. On the other hand, subsystems belonging to different coalitions, even though they are physically neighbors to each other, do not communicate, as in a decentralized scheme. Additionally, similar to the approach in [21], the scheme is time-varying. A flow diagram that summarizes the overall method is shown in Fig. 7.1.

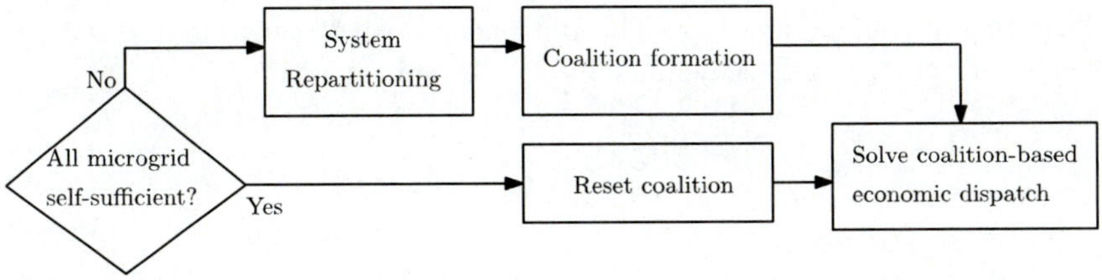

Fig. 7.1 (After [2]) The algorithmic flowchart of the proposed method at each repartitioning instant

7.2 Definitions

Prior to presenting the proposed methodology, some definitions are established. To that end, recall that $u^{\text{dg}}_{i,t}$ is the power that can be generated by the dispatchable generation units of node i and $d_{i,t}$ is the aggregated load of node i.

Definition 7.1 (*Local imbalance*) The local power imbalance of a subset of nodes $\mathcal{M} \subseteq \mathcal{N}$ at any $t \in \mathbb{Z}_{\geq 0}$, denoted by $\Delta^{\text{im}}_{\mathcal{M},t}$, is defined as

$$\Delta^{\text{im}}_{\mathcal{M},t} = \sum_{i \in \mathcal{M}} \left(-\mathbb{1}^{\top}_{n_i^{\text{dg}}} u^{\text{dg}}_{i,t} + d_{i,t} \right), \tag{7.1}$$

where $d_{i,t} = d^{\text{f}}_{i,t} + w^{\max}_{i,t}$ (see (2.6)) with $w^{\max}_{i,t}$ denotes the worst-case disturbance. □

In view of Definition 7.1, the local imbalance of subsystem $\mathcal{M}_{p,t}$ indicates the difference between the aggregated worst-case load of subsystem $\mathcal{M}_{p,t}$ and the local power generation of subsystem $\mathcal{M}_{p,t}$. Then, the self-sufficiency of a subsystem is defined as follows.

Definition 7.2 (*Self-sufficiency*) A subset of nodes $\mathcal{M} \subseteq \mathcal{N}$ at any $t \geq 0$ is self-sufficient for a period T if it has non-positive local imbalance at any step along the period $\mathcal{T} = \{t, t+1, \ldots, t+T-1\}$, i.e., $\Delta^{\text{im}}_{\mathcal{M},\tau} \leq 0$, for all $\tau \in \mathcal{T}$. □

Self-sufficiency can be suitably defined with the considered economic dispatch problem, which has a certain time horizon h, i.e., setting $T = h$. Note that, in the context of electrical energy systems, microgrids, which can be disconnected from the main grids or other microgrids (when in the island mode), must be self-sufficient. Definition 7.2 will be used as the criterion to decide whether a repartitioning process is necessary at time instant t. Moreover, using the notion of imbalance in Definition 7.1, the imbalance cost, which will be considered in the repartitioning problem, is defined as follows.

Definition 7.3 (*Imbalance cost*) The imbalance cost of subsystem $p \in \mathcal{P}$, with $\mathcal{M}_{p,t} \in \boldsymbol{\mathcal{M}}_t$, at any $t \in \mathbb{Z}_{\geq 0}$, denoted by $J^{\text{im}}_{p,t}$, is defined as

$$J^{\text{im}}_{p,t} = \sum_{\tau \in \mathcal{T}} \max\left(0, \Delta^{\text{im}}_{\mathcal{M}_{p,t},\tau}\right), \tag{7.2}$$

where $\Delta^{\text{im}}_{\mathcal{M}_{p,t},\tau}$ is defined based on (7.1). □

The imbalance cost (7.2) penalizes a subsystem that does not have enough local power resource to meet the loads over the whole prediction horizon. On the other hand, another repartitioning objective, namely the efficiency of each subsystem, is also considered and provided in the following definition.

Definition 7.4 (*Efficiency cost*) The efficiency cost of subsystem $p \in \mathcal{P}$, with $\mathcal{M}_{p,t} \in \boldsymbol{\mathcal{M}}_t$ at any $t \in \mathbb{Z}_{\geq 0}$, is defined as follows:

$$\begin{aligned} J^{\text{ef}}_{p,t} = \min_{\{(\boldsymbol{u}_{i,t},\boldsymbol{v}_{i,t}) \in \mathcal{L}_{i,t}\}_{i \in \mathcal{M}_{p,t}}} & \sum_{i \in \mathcal{M}_{p,t}} \left(f^{\text{l}}_i(\boldsymbol{u}_{i,t}) + f^{\text{c}}_i(\boldsymbol{v}_{i,t})\right) + f^{\text{e}}_p(\{\boldsymbol{v}_{i,t}\}_{i \in \mathcal{M}_{p,t}}) \\ \text{s.t. } & \boldsymbol{v}^j_{i,t} + \boldsymbol{v}^i_{j,t} = 0, \ \forall j \in \mathcal{N}_i \cap \mathcal{M}_{p,t}, \ \forall i \in \mathcal{M}_{p,t}, \end{aligned} \tag{7.3}$$

where $f^{\text{e}}_p : \mathbb{R}^{\sum_{i \in \mathcal{M}_{p,t}} h|\mathcal{N}_i|} \to \mathbb{R}$ is defined as follows:

$$f^{\text{e}}_p(\{\boldsymbol{v}_{i,t}\}_{i \in \mathcal{M}_{p,t}}) = \sum_{i \in \mathcal{M}_{p,t}} \sum_{j \in \mathcal{N}_i \setminus \mathcal{M}_{p,t}} c^{\text{et}}_i \|\boldsymbol{v}^j_{i,t}\|^2_2, \tag{7.4}$$

in which c^{et}_i is the extra per-unit cost of transferring power. □

The auxiliary cost function $f^{\text{e}}_p(\{\boldsymbol{v}_{i,t}\}_{i \in \mathcal{M}_{p,t}})$ adds an extra cost on the power transferred between one subsystem to another in order to minimize the dependency on the neighbors. Note that c^{et}_i can be set quite large to incentivize the decoupling among the subsystems. Moreover, as can be seen in Definition 7.4, in order to compute $J^{\text{ef}}_{p,t}$, the local controller must solve a local economic dispatch problem over h time instants, which is derived from (2.15), where the coupling constraints between neighboring subsystems are relaxed.

7.3 Online Repartitioning Scheme

Suppose that the system is allowed to be repartitioned at every t_{r} steps, where $t_{\text{r}} \geq 1$, i.e., the system may be repartitioned at $t \in \mathcal{T}_{\text{r}} = \{0, t_{\text{r}}, 2t_{\text{r}}, 3t_{\text{r}}, \ldots\}$, which is called the set of repartitioning instants. Therefore, the partitions remain the same at least until the next repartitioning instant, i.e., $\boldsymbol{\mathcal{M}}_t = \boldsymbol{\mathcal{M}}_{\lfloor t/t_{\text{r}} \rfloor t_{\text{r}}}$, for all $t \in \mathbb{Z}_{\geq 0}$. Based on this rule, unless specified otherwise, let $T = t_{\text{r}} + h$. Due to the time-varying nature of the loads, the self-sufficiency condition of each subsystem (c.f. Definition 7.2)

Algorithm 7.1 Repartitioning triggering mechanism

1. If $t \in \mathcal{T}_r$, for each subsystem $p \in \mathcal{P}$, with the set of nodes $\mathcal{M}_{p,t-1} \in \boldsymbol{\mathcal{M}}_{t-1}$, evaluate its self-sufficiency at t based on Definition 7.2, with $T = t_r + h$.
2. If a subsystem is not self-sufficient, raise a flag to start repartitioning procedure. Otherwise wait until all subsystems perform step 1.
3. If the flag to start the repartitioning procedure is not raised, then keep the current partitions, i.e., $\boldsymbol{\mathcal{M}}_t = \boldsymbol{\mathcal{M}}_{t-1}$.

might change over time. When a subsystem is not self-sufficient, then the repartitioning process is triggered. In this regard, the triggering mechanism is provided in Algorithm 7.1. At each time instant t, first the controllers in the network execute Algorithm 7.1. Furthermore, when the flag to perform repartitioning is raised, it is assumed that all other subsystems can receive this information. This assumption can be fulfilled if there is an either all-to-all communication network or, at worst, a connected communication network. When the communication network of the subsystems is connected, there exists a path from a subsystem to any other subsystem that can be used to relay this information.

Supposing that the network is triggered to repartition itself, the repartitioning problem that will be solved is stated next. First, assume that the network is initially partitioned into m non-overlapping subsystems and denote the set of initial partition at $t = 0$ by $\boldsymbol{\mathcal{M}}_0 = \{\mathcal{M}_{1,0}, \mathcal{M}_{2,0}, \ldots, \mathcal{M}_{m,0}\}$. Thus, for some time instant t, at which the system must perform repartitioning, the optimization problem that must be solved is as follows:

$$\underset{\boldsymbol{\mathcal{M}}_t}{\text{minimize}} \sum_{p=1}^{m} J^{\pi}(\mathcal{M}_{p,t}) \tag{7.5a}$$

$$\text{s.t.} \bigcup_{p=1}^{m} \mathcal{M}_{p,t} = \mathcal{N}, \quad \mathcal{M}_{p,t} \cap \mathcal{M}_{q,t} = \emptyset, \ \forall \mathcal{M}_{p,t}, \mathcal{M}_{q,t} \in \boldsymbol{\mathcal{M}}_t, \ p \neq q, \tag{7.5b}$$

$$\lambda_2(\mathcal{G}_{p,t}) > 0, \quad \forall \mathcal{M}_{p,t} \in \boldsymbol{\mathcal{M}}_t, \tag{7.5c}$$

$$\boldsymbol{\mathcal{M}}_t(0) = \boldsymbol{\mathcal{M}}_{t-1}. \tag{7.5d}$$

The cost function $J^{\pi}(\mathcal{M}_{p,t})$ is defined by

$$J^{\pi}(\mathcal{M}_{p,t}) = \alpha J^{\text{im}}_{p,t} + J^{\text{ef}}_{p,t}, \tag{7.6}$$

where α is the tuning parameter to determine the trade-off between both the imbalance and efficiency costs. Moreover, $\mathcal{G}_{p,t} = (\mathcal{M}_{p,t}, \mathcal{E}_{p,t})$ denotes the subgraph of subsystem p, with the set of edges denoted by $\mathcal{E}_{p,t} = \{(i, j) \in \mathcal{E} : i, j \in \mathcal{M}_{p,t}\}$ and $\lambda_2(\mathcal{G}_{p,t})$ denotes the second smallest eigenvalue of the Laplacian matrix of subgraph $\mathcal{G}_{p,t}$. Equation (7.5c) implies the requirement of having connected subsystems, i.e., the subgraph formed by each subsystem is connected. This constraint is imposed to avoid decoupling among the nodes within each subsystem. Furthermore, $\mathcal{M}_t(0)$ denotes the partition at the first iteration $k = 0$, which is obtained from the partition at the previous time instant, $t - 1$. In addition, Assumption 7.1, which is related to the initial partition $\mathcal{M}_0$, is considered.

Assumption 7.1 The initial partition $\mathcal{M}_0$ is non-overlapping with connected subsystems. □

Remark 7.1 For the case of electrical distribution networks, the initial partition $\mathcal{M}_0$ could be obtained by solving an optimal microgrid construction problem [22, 23]. □

Now, the repartitioning procedure, where the subsystems cooperatively solve Problem (7.5), is presented. To solve the repartitioning problem while taking into account the setup of the system, an iterative local improvement algorithm that is performed in a distributed and synchronous manner is proposed. The main idea of the algorithm is as follows. At each iteration, one node is proposed to be moved from one subsystem to a neighboring subsystem in order to improve the total cost. The algorithm is a variation of the Kernighan-Lin algorithm [8] and similar to the method presented in [24, 25]. However, differently from the proposed approach, the method in [24, 25] considers different cost function and constraints, e.g., balancing number of vertices. Additionally, unlike the method in [9, 10] and as explained later in Proposition 7.1, by considering moving only one vertex at each iteration, convergence to a local optimum can be shown.

Therefore, first consider the initial partition $\mathcal{M}_t(0)$. Moreover, denote the set of boundary nodes of subsystem $\mathcal{M}_{p,t}$, i.e., nodes that are connected (coupled) to at least one bus that belongs to another subsystem by $\mathcal{M}_{p,t}^{\mathrm{c}} = \{i : (i, j) \in \mathcal{E}, i \in \mathcal{M}_{p,t}, j \in \mathcal{N} \backslash \mathcal{M}_{p,t}\}$. Then, the repartitioning procedure is stated in Algorithm 7.2. The number of repartitioning iterations can be upper-bounded by a constant denoted by $k^{\max}$. Thus, the procedure is stopped when it reaches the maximum number of iteration $k^{\max}$. Proposition 7.1 characterizes the solution obtained by the proposed algorithm.

Proposition 7.1 *Let $\mathcal{M}_0$ be the initial partition at $t = 0$ and Assumption 7.1 hold. At any time instant at which the repartitioning process is triggered, the output of Algorithm 7.2 is a non-overlapping partition with connected subsystems and converges to a local minimum.* □

Proof The proof is the same as that of [2, Proposition 9], which is provided here for completeness.

Define by τ the time instant at which the repartitioning process is triggered, i.e., there exists at least one subsystem in $\mathcal{M}_\tau$ that is not self-sufficient. Let τ_0 be the first (smallest) repartitioning instant. Notice that the initial partition $\mathcal{M}_\tau(0)$, at any repartitioning instant τ, equals to the solution of Algorithm 7.2 at the previous repartitioning instant. Therefore, if at τ_0 the assertion holds, then it also holds for any repartitioning instants. Hence, it remains to evaluate the outcome of the repartitioning process at τ_0.

Algorithm 7.2 Repartitioning procedure

Suppose that subsystem $p \in \mathcal{P}$ is chosen randomly to propose a node that will be moved at the kth iteration. Then, the steps at each iteration are described below:

1. Subsystem p computes $J^{\pi}(\mathcal{M}_{p,t}(k))$, which is the local cost function at the kth iteration, based on (7.6).
2. Subsystem p computes a node that will be offered to be moved, denoted by $\theta_p(k)$ as follows:
$$\theta_p(k) \in \arg \min_{\theta \in \mathcal{M}^{\text{nd}}_{p,t}(k)} J^{\pi}(\mathcal{M}_{p,t}(k) \backslash \{\theta\}), \tag{7.7}$$
where $\mathcal{M}^{\text{nd}}_{p,t}(k) \subseteq \mathcal{M}^{\text{c}}_{p,t}(k)$ is a subset of boundary busses that do not disconnect subsystem p when removed, i.e., the graph form by $\mathcal{M}_{p,t}(k) \backslash \{\theta\}$, for $\theta \in \mathcal{M}^{\text{nd}}_{p,t}(k)$, is connected and with $T = t_{\text{r}} + h$ in the definition of J^{π}. The node $\theta_p(k)$ is randomly selected from the set of minimizers of (7.7).
3. Subsystem p computes the local cost difference if $\theta_p(k)$ is moved out from subsystem $\mathcal{M}_{p,t}$, i.e.,
$$\Delta J^{\pi}_p(k) = J^{\pi}(\mathcal{M}_{p,t}(k) \backslash \{\theta_p(k)\}) - J^{\pi}(\mathcal{M}_{p,t}(k)). \tag{7.8}$$
4. Subsystem p shares the information of $\theta_p(k)$ and $\Delta J^{\pi}_p(k)$ to the related neighboring subsystems $\phi(j, t) \in \mathcal{N}^{\text{s}}_p$, such that $\{\theta_p(k), j\} \in \mathcal{E}$.
5. All neighbors $q = \phi(j, t) \in \mathcal{N}^{\text{s}}_p$, such that $\{\theta_p(k), j\} \in \mathcal{E}$, compute the expected total cost difference if $\theta_p(k)$ is moved from subsystem p to subsystem q, as follows:
$$\Delta J^{\text{to}}_q(k) = J^{\pi}(\mathcal{M}_{q,t}(k) \cup \{\theta_p(k)\}) - J^{\pi}(\mathcal{M}_{q,t}(k)) + \Delta J^{\pi}_p(k), \tag{7.9}$$
and send the information of $\Delta J^{\text{to}}_q(k)$ to subsystem p.
6. Subsystem p selects the neighbor that will receive $\theta_p(k)$ as follows:
$$q^{\star} \in \arg \min_{q \in \mathcal{N}^{\text{s}}_p} \Delta J^{\text{to}}_q(k), \quad \text{s.t. } \{\theta_p(k), j\} \in \mathcal{E}, \; j \in \mathcal{M}_{q,t}(k), \tag{7.10}$$
where $q^{\star}$ is randomly chosen from the set of minimizers.
7. If $\Delta J^{\text{to}}_{q^{\star}}(k) \leq 0$, then the partition is updated as follows:
$$\mathcal{M}_{p,t}(k+1) = \mathcal{M}_{p,t}(k) \backslash \{\theta_p(k)\}, \tag{7.11}$$
$$\mathcal{M}_{q^{\star},t}(k+1) = \mathcal{M}_{q^{\star},t}(k) \cup \{\theta_p(k)\}. \tag{7.12}$$
Otherwise, the algorithm jumps to the next iteration, $k + 1$.

Since the system is not repartitioned when $t < \tau_0$, the initial partition at τ_0, $\boldsymbol{\mathcal{M}}_{\tau_0}(0) = \boldsymbol{\mathcal{M}}_0$, is non-overlapping with connected subsystems due to Assumption 7.1. Moreover, at any iteration of the repartitioning procedure, $k \leq k^{\max}$, the node proposed to be moved is selected from $\mathcal{M}^{\text{nd}}_{p,\tau_0}(k)$, which is the set of boundary nodes that do not cause the disconnection of the associated subsystem when removed (see (7.7)). At the end of the iteration, either one node is moved from one subsystem to another or no node is moved. These facts imply that, at the end of any iteration, $\boldsymbol{\mathcal{M}}_{\tau_0}(k)$ remains non-overlapping and the connectivity of each subsystem is maintained.

Now, the convergence of the repartitioning solution is shown. To this end, recall that the total cost at the beginning of iteration k is expressed by $\sum_{p\in\mathcal{P}} J^{\pi}(\mathcal{M}_{p,\tau}(k))$. The convergence is proved by showing that the evolution of the total cost is non-increasing. Suppose that $\theta_p(k)$ is moved from subsystem p to subsystem $q^{\star}$. Therefore, it follows that

$$\begin{aligned}
&\sum_{p\in\mathcal{P}} J^{\pi}(\mathcal{M}_{p,\tau}(k+1)) - \sum_{p\in\mathcal{P}} J^{\pi}(\mathcal{M}_{p,\tau}(k)) \\
&= J^{\pi}(\mathcal{M}_{p,\tau}(k+1)) - J^{\pi}(\mathcal{M}_{p,\tau}(k)) + J^{\pi}(\mathcal{M}_{q^{\star},\tau}(k+1)) - J^{\pi}(\mathcal{M}_{q^{\star},\tau}(k)) \\
&= J^{\pi}(\mathcal{M}_{p,\tau}(k)\backslash\{\theta_p(k)\}) - J^{\pi}(\mathcal{M}_{p,\tau}(k)) + J^{\pi}(\mathcal{M}_{q^{\star},\tau}(k) \cup \{\theta_p(k)\}) \\
&\quad - J^{\pi}(\mathcal{M}_{q^{\star},\tau}(k)) = \Delta J^{\mathrm{to}}_{q^{\star}}(k) \leq 0.
\end{aligned}$$

The first equality follows from the fact that only the local costs of subsystems p and $q^{\star}$ change after iteration k. The third equality follows from (7.8) and (7.9), and the inequality comes from the condition imposed in step 7 of Algorithm 7.2, where $\theta_p(k)$ is not moved if $\Delta J^{\mathrm{to}}_{\mathrm{t},q^{\star}}(k) > 0$. When no node is moved, the last inequality becomes an equality. □

Remark 7.2 After the network is repartitioned by Algorithm 7.2, not all subsystems might be self-sufficient. Note that in general, Problem (2.3) might actually have feasible solutions that require high power exchange, implying it might be impossible to partition the network into self-sufficient subsystems. □

7.4 Coalition-Based Economic Dispatch Scheme

In this section, the coalition-based economic dispatch scheme is discussed. Specifically, each self-sufficient subsystem must solve its local economic dispatch problem without exchanging power with its neighbors. Therefore, self-sufficient subsystems do not need to communicate with its neighbors to dispatch its components. However, for any subsystem that is not self-sufficient, its local economic dispatch problem might be infeasible since there is a deficiency of local power production in meeting the load. Since the repartitioning outcome does not guarantee the self-sufficiency of each subsystem, then the subsystems that are not self-sufficient must form a coalition with some other subsystems such that the resulting economic dispatch problem that must be solved by each coalition is feasible.

7.4.1 Coalition Formation

In a coalitional control scheme, only subsystems that belong to the same coalition can communicate among each other; thus, cooperatively compute their control inputs

[18, 19]. Therefore, it is necessary to construct the coalitions of subsystems that are relevant to the considered economic dispatch problem and this section presents a procedure to do so. To that end, denote by $\mathcal{C}_{p,t}$ and $\mathcal{D}_{p,t}$ the set of nodes and the set of subsystems that belong to the coalition of subsystem p, respectively. Then, assign one pair $(\mathcal{C}_{p,t}, \mathcal{D}_{p,t})$ to each subsystem $p \in \mathcal{P}$ to keep tracking the nodes and neighboring subsystems with which it forms a coalition. The coalition formation mechanism is described in Algorithm 7.3. The outcome of Algorithm 7.3 is stated in Proposition 7.2.

Algorithm 7.3 Coalition formation procedure

Initialization: Each subsystem $\mathcal{M}_{p,t}$ defines $\mathcal{C}_{p,t}(0) = \mathcal{M}_{p,t}$ and $\mathcal{D}_{p,t}(0) = \{p\}$.
Iteration: While $k < m - 1$, do:

1. Each subsystem p evaluates whether its coalition is self-sufficient based on Definition 7.2, i.e., whether
$$\Delta^{\text{im}}_{\mathcal{C}_{p,t}(k),\tau} \leq 0, \quad \forall \tau \in \{t, \ldots, t + t_{\text{r}} + h - 1\},$$
holds true.
2. If coalition $\mathcal{C}_{p,t}(k)$ is self-sufficient, then subsystem p waits until the iterations stop, i.e., $k = m - 1$.
3. Otherwise, subsystem p initiates a coalition merger by sending $\Delta^{\text{im}}_{\mathcal{C}_{p,t}(k),\tau}$, for all $\tau \in \mathcal{T}$, with $T = t_{\text{r}} + h$, to the subsystems that do not belong to coalition $\mathcal{C}_{p,t}(k)$ but they have physical connections with at least one node in coalition $\mathcal{C}_{p,t}(k)$, i.e., $q \in \mathcal{N}^{\text{c}}_{p,t} = \{q \in \mathcal{P} : \{i, j\} \in \mathcal{E}, i \in \mathcal{C}_{p,t}(k), j \in \mathcal{M}_{q,t}, \mathcal{C}_{q,t}(k) \neq \mathcal{C}_{p,t}(k)\}$. Note that if $\mathcal{N}^{\text{c}}_{p,t} = \emptyset$, then subsystem p does nothing until the iterations stop, i.e., $k = m - 1$.
4. For each neighbor $q \in \mathcal{N}^{\text{c}}_{p,t}$, if it is not communicating with another subsystem, then it computes
$$J^{\text{cim}}_q = \sum_{\tau \in \mathcal{T}} \max\left(0, \Delta^{\text{im}}_{\mathcal{C}_{q,t}(k),\tau} + \Delta^{\text{im}}_{\mathcal{C}_{p,t}(k),\tau}\right),$$
with $T = t_{\text{r}} + h$. Otherwise, $J^{\text{cim}}_q = \infty$. Then, it sends back J^{cim}_q to coalition $\mathcal{C}_{p,t}(k)$.
5. Based on J^{cim}_q, subsystem p chooses the neighbor with which it will merge as a coalition, as follows:
$$q^\star \in \arg\min_{q \in \mathcal{N}^{\text{c}}_{p,t}} J^{\text{cim}}_q \quad \text{s.t. } J^{\text{cim}}_q < \infty.$$
6. Update the coalition sets, i.e., $\mathcal{C}_{\rho,t}(k+1) = \mathcal{C}_{\rho,t}(k) \cup \mathcal{C}_{q^\star,t}(k)$ and $\mathcal{D}_{\rho,t}(k+1) = \mathcal{D}_{\rho,t}(k) \cup \mathcal{D}_{q^\star,t}(k)$, for all $\rho \in \mathcal{D}_{p,t}(k)$ and $\mathcal{C}_{\rho,t}(k+1) = \mathcal{C}_{\rho,t}(k) \cup \mathcal{C}_{p,t}(k)$ and $\mathcal{D}_{\rho,t}(k+1) = \mathcal{D}_{\rho,t}(k) \cup \mathcal{D}_{p,t}(k)$, for all $\rho \in \mathcal{D}_{q^\star,t}(k)$.
7. $k \leftarrow k + 1$ and go back to step 1.

Finally, for each subsystem p, set $\mathcal{C}_{p,\tau} = \mathcal{C}_{p,t}$, for $\tau = t + 1, \ldots, t + t_{\text{r}} - 1$.

Proposition 7.2 *The output coalitions of Algorithm 7.3, $\mathcal{C}_{p,\tau}$, for all $p \in \mathcal{P}$ and $\tau = t, \ldots, t + t_{\text{r}} - 1$, are self-sufficient for the period $T = h$.* □

Proof At each iteration $k < m - 1$, the evaluation in step 1 has two mutually exclusive outcomes:

1. All coalitions are self-sufficient.
2. There exist some coalitions that are not self-sufficient.

In case 1, it holds that $\mathcal{C}_{p,t}(m-1) = \mathcal{C}_{p,t}(k)$, for all $p \in \mathcal{P}$ since the coalitions do not change from the kth iteration until the $(m-1)$th iteration. Note that when all subsystems $\mathcal{M}_{p,t} \in \boldsymbol{\mathcal{M}}_t$ are self-sufficient, then $\mathcal{C}_{p,t}(0)$, for all $p \in \mathcal{P}$, are self-sufficient. Therefore, this case is also included here. Since in Step 1, the local imbalance of each coalition $\mathcal{C}_{p,t}$ is assessed from t until $t + t_\text{r} + h - 1$ steps, self sufficiency for the period $T = h$, for each coalition $\mathcal{C}_{p,\tau}$, $\tau = t, \ldots, t + t_\text{r} - 1$, is guaranteed. In case 2, according to steps 3–6 in Algorithm 7.3, at least one of the coalitions that are not self-sufficient will be merged with one of its neighboring coalitions at the next iteration $k + 1$. Since the number of initial coalitions is finite (m), then if case 2 keeps occurring, all coalitions will be merged into one grand coalition, i.e., $\mathcal{C}_{p,t} = \mathcal{N}$, for all $p = 1, \ldots, m$, at a finite number of iterations. For this case, due to Assumption 2.3, the grand coalition is always self-sufficient. Otherwise, case 1 will occur. Furthermore, in case 2, the minimum number of coalitions that can perform steps 3–6 (merging with one of its neighboring coalitions) is one. If, for $k \geq 1$, only one coalition merges with one of its neighbors, then it requires $m - 1$ iterations to merge all coalitions. Note that if more than one pair of coalitions merges at least in one iteration, then the number of iterations required to merge all of them is less than $m - 1$. □

Remark 7.3 Notice that in steps 3–6 of Algorithm 7.3 more than one coalition that is not self-sufficient can initiate a coalition merger. However, in step 4, each coalition can only be asked by one neighbor at each iteration. In this regard, step 4 can be executed by the principle of first comes first served. □

7.4.2 Coalition-Based Economic Dispatch

This section discusses a non-centralized scheme to solve Problem (2.3) based on the coalitions that have been formed. Note that when all subsystems $p \in \mathcal{P}$ are self-sufficient, the coalitions are reset as in the initialization of Algorithm 7.3, i.e., $\mathcal{C}_{p,t} = \mathcal{M}_{p,t}$, for all $p \in \mathcal{P}$. First, Problem (2.3) is reformulated as a coalition-based problem and the reformulation is shown in Proposition 7.3.

Proposition 7.3 *Suppose that, at time instant t, the network is partitioned into m non-overlapping subsystems, defined by the set $\boldsymbol{\mathcal{M}}_t = \{\mathcal{M}_{p,t} : p \in \mathcal{P}\}$. Furthermore, coalitions of subsystems, denoted by $\mathcal{C}_{p,t}$, for all $p \in \mathcal{P}$, are formed based on Algorithm 7.3. Then, Problem (2.3) is equivalent to*

$$\underset{\{\boldsymbol{u}_{i,t},\boldsymbol{v}_{i,t}\}_{i\in\mathcal{N}}}{\text{minimize}} \sum_{p\in\mathcal{P}} \sum_{i\in\mathcal{M}_{p,t}} \left(f_i^{\mathrm{l}}(\boldsymbol{u}_{i,t}) + f_i^{\mathrm{c}}(\boldsymbol{v}_{i,t})\right) \tag{7.13a}$$

$$\text{s.t. } (\boldsymbol{u}_{i,t}, \boldsymbol{v}_{i,t}) \in \mathcal{L}_{i,t}, \ \forall i \in \mathcal{C}_{p,t}, \ \forall p \in \mathcal{P}, \tag{7.13b}$$

$$\boldsymbol{v}_{i,t}^{j} + \boldsymbol{v}_{j,t}^{i} = 0, \quad \forall j \in \mathcal{N}_i \cap \mathcal{C}_{p,t}, \ \forall i \in \mathcal{C}_{p,t}, \ \forall p \in \mathcal{P}, \tag{7.13c}$$

$$\boldsymbol{v}_{i,t}^{j} + \boldsymbol{v}_{j,t}^{i} = 0, \quad \forall j \in \mathcal{N}_i \backslash \mathcal{C}_{p,t}, \ \forall i \in \mathcal{C}_{p,t}, \ \forall p \in \mathcal{P}. \tag{7.13d}$$

Proof Notice that $\bigcup_{p\in\mathcal{P}} \mathcal{C}_{p,t} = \mathcal{N}$. Then, the proof follows the proof of Proposition 2.1. □

Remark 7.4 For each coalition $\mathcal{C}_{p,t}$, (7.13b) and (7.13c) are local constraints. Spefically for the constraints in (7.13c), some of them might involve two different subsystems. On the other hand, (7.13d) couples two different coalitions. □

Suppose that at the end of the coalition formation procedure there exist c distinct coalitions whose elements are different from one to another, where $c \leq m$. Note that when $\mathcal{C}_{p,t} = \mathcal{C}_{q,t}$, subsystems p and q belong to the same coalition. A non-centralized economic dispatch scheme will be formulated for these coalitions by decomposing Problem (7.13) such that each coalition solves its own economic dispatch. The decomposition is done by not allowing power exchange between two neighboring coalitions. Thus, the MPC-based economic dispatch problem must be solved at each coalition $\mathcal{C}_{p,t}$, for all $p \in \mathcal{P}$, is stated as follows:

$$\underset{\{\boldsymbol{u}_{i,t},\boldsymbol{v}_{i,t}\}_{i\in\mathcal{C}_{p,t}}}{\text{minimize}} \sum_{i\in\mathcal{C}_{p,t}} \left(f_i^{\mathrm{l}}(\boldsymbol{u}_{i,t}) + f_i^{\mathrm{c}}(\boldsymbol{v}_{i,t})\right) \tag{7.14a}$$

$$\text{s.t. } (\boldsymbol{u}_{i,t}, \boldsymbol{v}_{i,t}) \in \mathcal{L}_{i,t}, \ \forall i \in \mathcal{C}_{p,t}, \tag{7.14b}$$

$$\boldsymbol{v}_{i,t}^{j} + \boldsymbol{v}_{j,t}^{i} = 0, \quad \forall j \in \mathcal{N}_i \cap \mathcal{C}_{p,t}, \ \forall i \in \mathcal{C}_{p,t}, \tag{7.14c}$$

$$\boldsymbol{v}_{i,t}^{j} = 0, \quad \forall j \in \mathcal{N}_i \backslash \mathcal{C}_{p,t}, \ \forall i \in \mathcal{C}_{p,t}. \tag{7.14d}$$

Note that if subsystems p and q belong to the same coalition at time step t, i.e., $\mathcal{C}_{p,t} = \mathcal{C}_{q,t}$, then they must cooperatively solve the same problem in a distributed manner. Additionally, if $c = m$, then each subsystem is self-sufficient, implying a fully decentralized scheme is applied to the network. On the other hand, if $c = 1$, then a fully distributed scheme is applied to the network.

Now, it is shown that Problem (7.14), for any coalition, has a solution. Furthermore, the solution to Problem (7.14) is also a feasible solution to the original problem (2.3).

Proposition 7.4 *Let the coalitions $\mathcal{C}_{p,t}$, for all $p \in \mathcal{P}$, are formed by using Algorithm 7.3. Then, there exists a non-empty set of optimal solutions to Problem (7.14), for each coalition $\mathcal{C}_{p,t}$, where $p \in \mathcal{P}$, and any $t \in \mathbb{Z}_{\geq 0}$.* □

Proof Since the cost function is convex and the constraints form a compact and convex set, the existence of optimal solutions is guaranteed provided that the feasible set is nonempty. Therefore, it remains to show that Problem (7.14), for any $\mathcal{C}_{p,t}$, has a non-empty feasible set. If Algorithm 7.3 results in one coalition over the whole network, i.e., $\mathcal{C}_{p,t} = \mathcal{N}$, for $p = 1, \ldots, m$, then it implies that all subsystems must solve the centralized economic dispatch problem (7.13) cooperatively. Therefore, in this case, for any $\mathcal{C}_{p,t}$, Problem (7.14) is equal to Problem (7.13). Due to Assumption 2.3, feasible solutions to Problem (7.13) exist. Otherwise, Algorithm 7.3 results in at least two different self-sufficient coalitions. In this case, each coalition is self-sufficient for the period $T = h$ (see Proposition 7.2), i..e, it has non-positive local imbalance for the whole prediction horizon $\{t, \ldots, t + h - 1\}$. Based on (7.1), the worst-case uncertain imbalance between loads and non-dispatchable generation can be met cooperatively by the distributed generation units within the coalition. Therefore, there exists a non-empty subset of feasible solution of Problem (7.13) such that (7.14d), for each $\mathcal{C}_{p,t}$, where $p = 1, \ldots, m$, holds, implying the existence of a non-empty feasible set of Problem (7.14). □

Proposition 7.5 *Let $(\boldsymbol{u}^\star_{i,t}, \boldsymbol{v}^\star_{i,t})$, for all $i \in \mathcal{C}_{p,t}$, be a solution to Problem (7.14), for all coalitions $\mathcal{C}_{p,t}$, where $p \in \mathcal{P}$. Then, they collectively are also a feasible solution to Problem (2.3).* □

Proof By Proposition 7.3, Problem (7.13) is equivalent to Problem (2.3). Therefore, it remains to show that $(\boldsymbol{u}^\star_{i,\tau}, \boldsymbol{v}^\star_{i,\tau})$, for all $i \in \mathcal{N}$, is a feasible solution to Problem (7.13). Note that Problem (7.14) is obtained by decomposing Problem (7.13). As can be seen, the constraints (7.13b)–(7.13c) are decomposed for each coalition and considered as (7.14b)-(7.14c) in Problem (7.14). Since $(\boldsymbol{u}^\star_{i,t}, \boldsymbol{v}^\star_{i,t})$, for all $i \in \mathcal{C}_{p,t}$, satisfy the constraints (7.14b)–(7.14c), they also satisfy (7.13b)–(7.13c). Finally, for any $\mathcal{C}_{p,t}$, by (7.14d), it holds that $\boldsymbol{v}^{j\star}_{i,t} = \boldsymbol{v}^{i\star}_{j,t} = 0$, for all $j \in \mathcal{N}_i \backslash \mathcal{C}_{p,t}$ and $i \in \mathcal{C}_{p,t}$. From this fact, it follows that $\boldsymbol{v}^{j\star}_{i,t} + \boldsymbol{v}^{i\star}_{j,t} = 0$ for all $j \in \mathcal{N}_i \backslash \mathcal{C}_{p,t}$ and $i \in \mathcal{C}_{p,t}$, implying the satisfaction of the constraints in (7.13d). □

Remark 7.5 The main challenge in distributedly solving Problem (7.14) is the existence of coupling constraints among the subsystems in the same coalition, i.e.,

$$\boldsymbol{v}^j_{i,t} + \boldsymbol{v}^i_{j,t} = 0, \quad \forall j \in \mathcal{N}_i \cap \mathcal{C}_{p,t} \backslash \mathcal{M}_{p,t}, \; \forall i \in \mathcal{C}_{p,t}. \tag{7.15}$$

One way to overcome the coupling constraints is by implementing the distributed methods presented in Chaps. 3 and 5. □

7.5 Suboptimality and Communication Complexity

This section discusses the suboptimality and communication burden of the proposed scheme. First, an estimation of the suboptimality level achieved by the scheme is shown. To that end, state the collection of the optimization problems (7.14), for all coalitions $\mathcal{C}_{p,t},\ p \in \mathcal{P}$, as follows:

$$\begin{aligned} &\underset{\{(\boldsymbol{u}_{i,t}, \boldsymbol{v}_{i,t})\}_{i \in \mathcal{N}}}{\text{minimize}} \sum_{i \in \mathcal{N}} \left(f_i^{\mathrm{l}}(\boldsymbol{u}_{i,t}) + f_i^{\mathrm{c}}(\boldsymbol{v}_{i,t}) \right) \\ &\text{s.t. (7.14b), (7.14c), and (7.14d)}, \forall p \in \mathcal{P}. \end{aligned} \tag{7.16}$$

Denote the optimal value of Problem (7.16) by $J_t^\star$. Note that $J_t^\star$ represents the cost function value of Problem (2.3) computed by the proposed scheme. Furthermore, denote by J^o the optimal value of Problem (2.3) and define the suboptimality measure as the difference between the cost function value computed using the proposed scheme and the optimal value of Problem (2.3), denoted by ΔJ_t, i.e.,

$$\Delta J_t = J_t^\star - J_t^o. \tag{7.17}$$

Therefore, an estimate of ΔJ_t is shown in Proposition 7.6.

Proposition 7.6 *Let $J_t^\star$ and J_t^o be the optimal values of Problems (7.16) and (2.3) at time t, respectively. Furthermore, let J_t^b denote the optimal value of the following optimization problem:*

$$\begin{aligned} &\underset{\{(\boldsymbol{u}_{i,t}, \boldsymbol{v}_{i,t})\}_{i \in \mathcal{N}}}{\text{minimize}} \sum_{i \in \mathcal{N}} \left(f_i^{\mathrm{l}}(\boldsymbol{u}_{i,t}) + f_i^{\mathrm{c}}(\boldsymbol{v}_{i,t}) \right) \\ &\text{s.t. } (7.14b) \text{ and } (7.14c),\ \forall p \in \mathcal{P}. \end{aligned} \tag{7.18}$$

Then, the following estimate on the suboptimality measure ΔJ_t, defined in (7.17), holds:

$$\Delta J_t \leq J_t^\star - J_t^b. \tag{7.19}$$

Proof Note that Problem (7.18) can be obtained by relaxing Problem (2.3). In particular, the coupling constraint $\boldsymbol{v}_{i,t}^j + \boldsymbol{v}_{j,t}^i = 0$, for each pair of nodes i and j that do not belong to the same coalition, is discarded in Problem (7.18). Due to this relaxation, it can be concluded that $J_t^b \leq J_t^o$. Moreover, based on Proposition 7.5, the solution obtained by Problem (7.14), for all coalitions $\mathcal{C}_{p,t},\ p \in \mathcal{P}$, is also a feasible solution to Problem (2.3), implying that $J_t^o \leq J_t^\star$. Based on the preceding observations, the relation in (7.19) holds. □

Remark 7.6 Consider the case when $\mathcal{C}_{p,t} = \mathcal{N}$, for all $p \in \mathcal{P}$. In this case, for any $i \in \mathcal{N}$, all neighbors of node i, i.e., $j \in \mathcal{N}_i$, belong to the same coalition as that of node i. Thus, in (7.14d), $\mathcal{N}_i \backslash \mathcal{C}_{p,t} = \emptyset$. This fact implies that Problem (7.16) is equivalent to Problem (2.3) and Problem (7.18), implying $\Delta J_t = 0$ and $J_t^\star - J_t^b = 0$. □

Remark 7.7 The optimal value of Problem (7.18) can be regarded as a lower bound of Problem (2.3). Problem (7.18) can also be decomposed into m sub-problems, each of which can be assigned to each coalition. Specifically, each sub-problem is stated as follows:

$$\begin{aligned} \underset{\{(\boldsymbol{u}_{i,t},\boldsymbol{v}_{i,t})\}_{i\in\mathcal{C}_{p,t}}}{\text{minimize}} \quad & \sum_{i\in\mathcal{C}_{p,t}} \left(f_i^{\text{l}}(\boldsymbol{u}_{i,t}) + f_i^{\text{c}}(\boldsymbol{v}_{i,t})\right) \\ \text{s.t. } & (7.14\text{b}) \text{ and } (7.14\text{c}) \end{aligned} \tag{7.20}$$

As mentioned in Remark 7.6, $\Delta J_t = J_t^\star - J_t^b = 0$ when $\mathcal{C}_{p,t} = \mathcal{N}$, for all $p \in \mathcal{P}$. Then, in this particular case, J_t^b is not necessary to be computed. □

Now, the communication cost of the proposed scheme is discussed. Algorithms 7.2 and 7.3 do require information exchange among the controllers. The total size of data exchanged throughout the process in Algorithm 7.2 is $\mathcal{O}(m)$ per iteration. It is obtained since, at each iteration, the subsystem selected to propose a node to be moved must send the information about the node, which is a scalar, to its neighbors and receives back the cost adjustment, which is also a scalar. Furthermore, the total size of information exchanged in Algorithm 7.3 is also $\mathcal{O}(m)$, since the information exchange process is similar to that of Algorithm 7.2. Finally, the size of data communicated when solving the coalition-based economic dispatch problem is evaluated. Each coalition might need to use a distributed optimization method since there might be more than one subsystem in a coalition. As an example, consider Algorithm 3.1 with one round communication as the distributed optimization method. In this algorithm, the size of exchanged information is $\mathcal{O}(m|\mathcal{N}|h)$ per iteration since each subsystem must exchange the coupled decision variables with each neighbor.

In the best-case scenario, communication might not be necessary at one time instant, particularly when all subsystems are self-sufficient. Furthermore, even if the repartitioning procedure is triggered, in the worst-case scenario, i.e., when the resulting coalition includes all subsystems, the extra amount of data must be exchanged to perform the repartitioning and coalition formation procedures is relatively much smaller than that of performing the distributed algorithm. In addition, for a coalition that only has one subsystem, its controller only needs to solve a local optimization problem once, which also significantly reduces the computational burden.

As stated in Remark 7.6, an optimal solution is only obtained when the scheme is fully distributed, i.e., all controllers belong to the same coalition. In other cases, the solution might be suboptimal. However, in those cases, the communication burden of the scheme is lower than when the scheme is fully distributed since some agents do not need to communicate to its neighbors that do not belong to the same coalition. Therefore, there is a trade-off between communication burden and suboptimality.

Finally, it is worth noting some points regarding the practicality of performing the proposed scheme. As in any distributed scheme, local controllers must cooperate to perform the scheme and a communication network must also be available. Since the partition of the physical network is time-varying, a dynamic communication network, containing necessary links, might be required. Another possibility is by having an all-to-all network, although in the process, not all links will be used. Furthermore, each local controller must also be able to communicate with the dispatchable components of the network, i.e., the storage and dispatchable generation units. The second important note is that although this chapter considers an MPC-based framework, where the set-points are computed at each time instant, the proposed method can also be implemented for a day-ahead economic dispatch without requiring any modification. In this case, the prediction horizon is set to be one day. On each day and prior to the computation of the decisions, the self-sufficiency of each subsystem is evaluated.

7.6 Numerical Study

Numerical study in this chapter considers the benchmark case (see Sect. 2.4) where the dispatchable, solar-based distributed generation, and storage units are placed according to Fig. 7.2. Moreover, the load profiles are generated based on the typical residential load profile whereas the non-dispatchable generation units are solar-powered. It is assumed that the subsystems have perfect knowledge of the loads and non-dispatchable generation units, i.e., $\boldsymbol{w}_{i,t}^{\max} = 0$, for all $t \geq \mathbb{Z}_{\geq 0}$. Note that, quadratic cost functions as defined in (2.16) and (2.18) are considered and the local constraint sets $\mathcal{U}_i$ and $\mathcal{V}_i$, for all $i \in \mathcal{N}$, considered in this simulation study are formed by (2.6)–(2.14). The simulation time is one day with the sampling time of 15 min, implying 96 time steps. Furthermore, the prediction horizon is set to be 8 time steps,

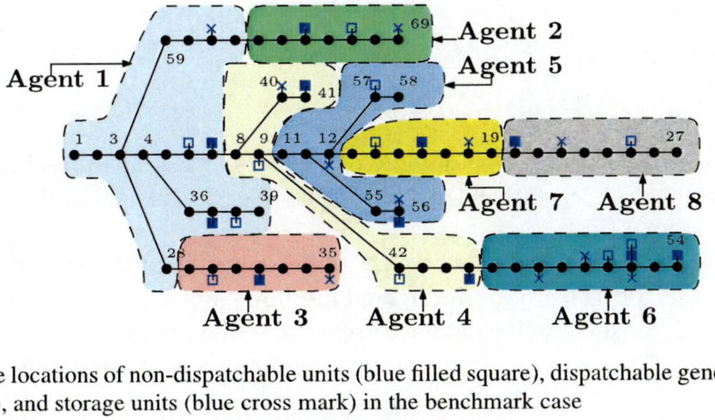

Fig. 7.2 The locations of non-dispatchable units (blue filled square), dispatchable generators (blue open square), and storage units (blue cross mark) in the benchmark case

Table 7.1 Parameters of the network components

Parameters	Value	Unit	Bus
$u_i^{\text{dg,min}}, u_i^{\text{dg,max}}$	0, 350	kW	$i \in \mathcal{N}^{\text{dg}}$
$x_i^{\min}, x_i^{\max}, x_{i,0}$	30, 100, 50	%	$i \in \mathcal{N}^{\text{st}}$
$u_i^{\text{ch}}, u_i^{\text{dh}}$	100, 100	kW	$i \in \mathcal{N}^{\text{st}}$
$e_{\text{cap},i}$	1000	kWh	$i \in \mathcal{N}^{\text{st}}$
a_i	1	–	$i \in \mathcal{N}^{\text{st}}$
$c_i^{\text{st}}, c_i^{\text{dg}}$	1, 10	–	$i \in \mathcal{N}$
$c_i^{\text{tp}}, c_i^{\text{t}}$	10, 1	–	$i \in \mathcal{N}$

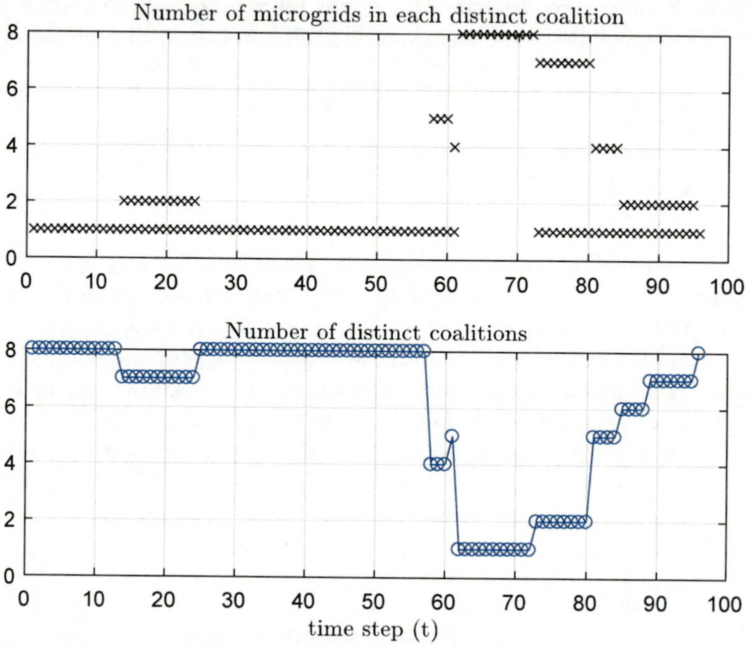

Fig. 7.3 The evolution of coalitions formed

the repartitioning period $t_r = 1$, and the parameters of the repartitioning procedure are set as follows $k^{\max} = 80$ and $\alpha = 10^4$. The other parameters of the components are given in Table 7.1.

The initial partition of the network is given in Sect. 2.4 and shown in Fig. 7.2. How the subsystems form coalitions throughout the simulation can be seen in Fig. 7.3. At $t = 1, \ldots, 13$, all subsystems obtained from the initial partition of the network are self-sufficient. Then, at $t = 14$, for the first time the network must be repartitioned. Moreover, for $t = 14, \ldots, 24$, the repartitioning and coalition formation procedures are always performed. During this period, two subsystems are not self-sufficient and

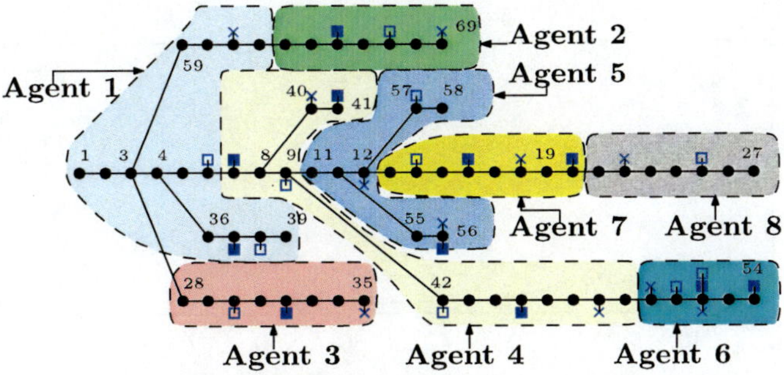

a. The partitioning result at $t = 25$.

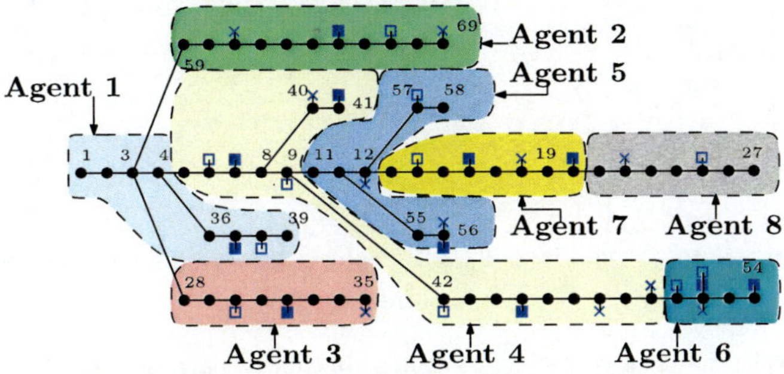

b. The partitioning result at $t = 96$.

Fig. 7.4 Partitioning results at **a** $t = 25$ and **b** $t = 96$. At both time instants, the resulting subsystems are self-sufficient

join together as a coalition. At $t = 25$, the repartitioning procedure produces self-sufficient subsystems, as shown in Fig. 7.4a and self-sufficiency is maintained until $t = 57$. However, during the peak hours, coalitions must be formed and even at some period all subsystems must join as one coalition. Gradual changes of the coalitions formed particularly at $t \in \mathbb{Z}_{\geq 57}$ can also be observed. Moreover, towards the end of the simulation it can also be seen that by performing repartitioning of the network, the number of self-sufficient subsystems improves as the number of distinct coalitions also increases. Finally, at $t = 96$, all the subsystems formed, as shown in Fig. 7.4b, are self-sufficient.

Figure 7.5 shows the cost values for all time instant and the suboptimality of the proposed scheme. As provided by Proposition 7.6, the controllers might also compute an upper bound of the suboptimality, which is shown by the dashed line in the bottom plot of Fig. 7.5. The average suboptimality throughout the simulation is 21.93%, whereas the average upper bound of the suboptimality defined in Proposition 7.6 is 51.38%. Furthermore, during $t = 62, \ldots, 72$, when all subsystems form one coalition, the optimal cost values are obtained since a fully distributed scheme is employed.

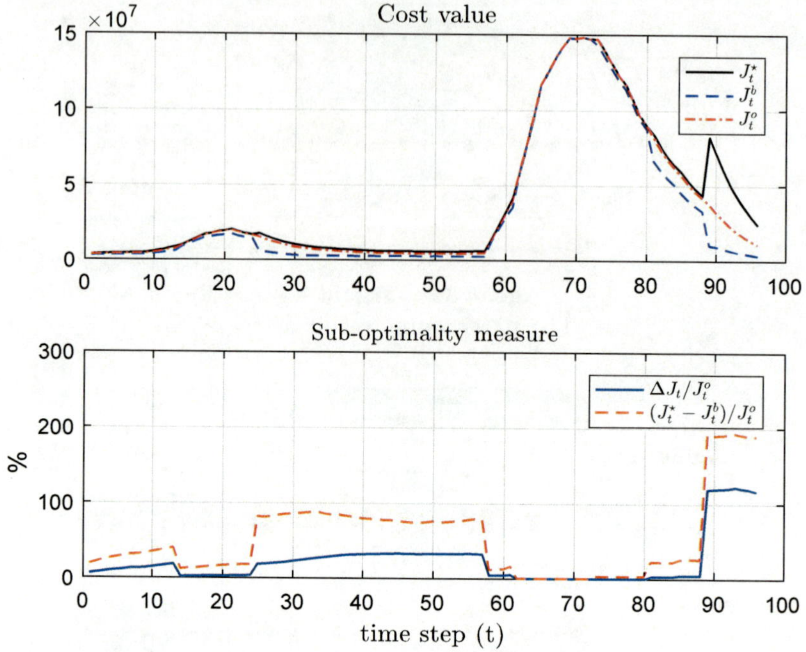

Fig. 7.5 Top plot shows the cost values computed using the proposed scheme, $J_t^\star$, (solid line), by solving Problem (2.3) centrally as the benchmark J_t^o (dashed-dotted line), and the lower bound, J_t^b (dashed line). Bottom plot shows the suboptimality of the proposed scheme and its upper bound

7.7 Summary

This chapter presents a novel non-centralized MPC-based economic dispatch scheme. The approach consists of an event-triggered repartitioning method with the aim of maintaining self-sufficiency of each subsystem and decomposing the centralized economic dispatch problem into coalition-based sub-problems in order to compute a feasible but possibly suboptimal decisions. The main advantage of the approach is a low communication burden, which is essential for online applications. Additionally, the effectiveness of the approach is also showcased in a numerical study. This chapter completes the discussion of this thesis on how to solve the economic dispatch problem in a non-centralized manner. Then, the next chapter discusses the second step of the MPC-based scheme, where the decisions that have been computed are implemented. Particularly, a cooperation issue, when not all subsystems comply with the computed decisions, is presented and a methodology to deal with this issue is developed.

References

1. Ananduta W, Ocampo-Martinez C (2019) Decentralized energy management of power networks with distributed generation using periodical self-sufficient repartitioning approach. In: Proceedings of the American control conference (ACC), pp 3230–3235
2. Ananduta W, Ocampo-Martinez C (2021) Event-triggered partitioning for non-centralized predictive-control-based economic dispatch of interconnected microgrids. Automatica (in press). Available at: https://doi.org/10.1016/j.automatica.2021.109829
3. Baker K, Guo J, Hug G, Li X (2016) Distributed MPC for efficient coordination of storage and renewable energy sources across control areas. IEEE Trans Smart Grid 7(2):992–1001
4. Wang T, O'Neill D, Kamath H (2015) Dynamic control and optimization of distributed energy resources in a microgrid. IEEE Trans Smart Grid 6(6):2884–2894
5. Kraning M, Chu E, Lavaei J, Boyd S (2014) Dynamic network energy management via proximal message passing. Found Trends Optim 1(2):73–126
6. Braun P, Grüne L, Kellett CM, Weller SR, Worthmann K (2016) A distributed optimization algorithm for the predictive control of smart grids. IEEE Trans Autom Control 61(12):3898–3911
7. Ananduta W, Pippia T, Ocampo-Martinez C, Sijs J, De Schutter B (2019) Online partitioning method for decentralized control of linear switching large-scale systems. J Franklin Inst 356(6):3290–3313
8. Kernighan BW, Lin S (1970) An efficient heuristic procedure for partitioning graphs. Bell Syst Tech J 49(2):291–307
9. Barreiro-Gomez J, Ocampo-Martinez C, Quijano N (2017) Partitioning for large-scale systems: a sequential distributed MPC design. IFAC-PapersOnLine 50(1):8838–8843
10. Barreiro-Gomez J, Ocampo-Martinez C, Quijano N (2019) Time-varying partitioning for predictive control design: density-games approach. J Process Control 75:1–14
11. Fjällström (1998) Algorithms for graph partitioning: a survey. Linköping Electron Articles Comput Inf Sci 3(10)
12. Gupta A (1997) Fast and effective algorithms for graph partitioning and sparse-matrix ordering. IBM J Res Dev 41(1–2):171–183
13. Nayeripour M, Fallahzadeh-Abarghouei H, Waffenschmidt E, Hasanvand S (2016) Coordinated online voltage management of distributed generation using network partitioning. Electr Power Syst Res 141:202–209
14. Guo J, Hug G, Tonguz OK (2016) Intelligent partitioning in distributed optimization of electric power systems. IEEE Trans Smart Grid 7(3):1249–1258
15. Ocampo-Martinez C, Bovo S, Puig V (2011) Partitioning approach oriented to the decentralised predictive control of large-scale systems. J Process Control 21(5):775–786
16. Maestre JM, Muñoz de la Peña D, Jiménez Losada A, Algaba E, Camacho EF (2014) A coalitional control scheme with applications to cooperative game theory. Optim Control Appl Methods 35(5):592–608
17. Fele F, Maestre JM, Mehdy Hashemy S, Muñoz de la Peña D, Camacho EF (2014) Coalitional model predictive control of an irrigation canal. J Process Control 24(4):314–325
18. Fele F, Maestre JM, Camacho EF (2017) Coalitional control: cooperative game theory and control. IEEE Control Syst Mag 37(1):53–69
19. Fele F, Debada E, Maestre JM, Camacho EF (2018) Coalitional control for self-organizing agents. IEEE Trans Autom Control 63(9):2883–2897
20. Muros FJ, Maestre JM, Algaba E, Alamo T, Camacho EF (2017) Networked control design for coalitional schemes using game-theoretic methods. Automatica 78:320–332
21. Núñez A, Ocampo-Martinez C, Maestre JM, De Schutter B (2015) Time-varying scheme for noncentralized model predictive control of large-scale systems. Math Probl Eng 2015:1–17
22. Barani M, Aghaei J, Akbari MA, Niknam T, Farahmand H, Korpås M (2018) Optimal partitioning of smart distribution systems into supply-sufficient microgrids. IEEE Trans Smart Grid (in press)

23. Arefifar SA, Mohamed YA-RI, El-Fouly THM (2012) Supply-adequacy-based optimal construction of microgrids in smart distribution systems. IEEE Trans Smart Grid 3(3):1491–1502
24. Gilbert JR, Zmijewski E (1987) A parallel graph partitioning algorithm for a message-passing multiprocessor. Int J Parallel Progr 16(6):427–449
25. Karypis G, Kumar V (1998) A fast and high quality multilevel scheme for partitioning irregular graphs. SIAM J Sci Comput 20(1):359–392

Chapter 8
Resiliency of Non-centralized MPC Schemes Against Adversaries

Non-centralized MPC-based economic dispatch schemes for energy systems rely on the cooperation of all agents. This chapter discusses the case in which some of the agents perform one type of adversarial actions (attacks) and they do not comply with the decisions computed by performing a non-centralized MPC algorithm. Moreover, this chapter specifically discusses energy systems whose subsystems behave as a microgrid, which can operate in either connected or island mode. A novel resilient non-centralized MPC scheme for such systems that can cope with non-compliance issue is proposed in this chapter. The approach consists of passive and active methods. The passive method robustifies the decisions, whereas the active method isolates the adversarial agents. Therefore, first, Sect. 8.1 provides an introduction of the non-compliance problem and an outlook of the proposed methodology. Then, Sect. 8.2 gives the adversary model considered in the system. Afterward, the passive and active methods are presented in Sects. 8.3 and 8.4, respectively. The combination of these methods is then shown in Sect. 8.5. Furthermore, the analysis of the proposed methodology is then given in Sect. 8.6, whereas the effectiveness of the methodology is showcased by numerical simulations in Sect. 8.7. Finally, Sect. 8.8 provides a summary of this chapter.

8.1 Non-compliance in Non-centralized Schemes

As discussed in Sect. 2.3, one of the main requirements for implementing a distributed approach is the cooperation among all agents, which must operate in compliance with the algorithm. However, it might happen that some agents in the network do not cooperate because they selfishly want to have a better performance or suffer from failures. This chapter discusses the case in which there are agents that might perform a certain adversarial behavior. In particular, it is considered that the adversarial agents do not comply with the decisions that are obtained from the distributed

W. W. Ananduta, *Non-centralized Optimization-Based Control Schemes for Large-Scale Energy Systems*, Springer Theses,
https://doi.org/10.1007/978-3-030-89803-8_8

algorithm implemented in the network. In other words, the adversarial agents might implement a different decision/control input than the one that is computed by using the distributed algorithm.

Non-compliance of some agents in a network that applies a distributed control approach has been discussed in some papers, e.g., [1, 2]. For instance, a secure dual-decomposition-based DMPC, in which each agent should monitor two neighbors that provide extreme control input values and disregard these extreme values, has been proposed in [1]. Furthermore, a cyber-attack problem of a consensus-based distributed control scheme for distributed energy storage systems has also been addressed in [2], where the approach involves a fuzzy-logic-based detection and a consensus-based leader-follower distributed control scheme. Related to the cyber-security issue of cyber-physical systems, the work presented in [3] provides a mathematical framework for attack detection and monitoring, particularly for deterministic systems. Moreover, consensus problems in which some of the agents perform an adversarial behavior to prevent convergence have also been investigated in [4–6]. In the DMPC framework, the issue that some agents might provide false information, which is a different type of adversarial behaviors, has also been discussed [7, 8]. A scenario-based defense mechanism [7] and a compensation scheme to incentivize truth telling among agents [8] have been proposed to deal with false information problems.

The main contribution of this chapter is the resilient method for non-centralized economic dispatch schemes of energy systems. The proposed approach combines both passive and active mechanisms to deal with the issue of unexpected disturbances. Specifically, the economic dispatch problem, which takes into account the adversarial behavior, is formulated as a chance-constrained problem in which attacks and loads are regarded as uncertain disturbances. In order to solve the problem, a stochastic two-step approach is applied. The first step of the approach is to compute probabilistic bounds using a randomization-based program while the second step is to solve a robust program that takes into account the probabilistic bounds. As a result, it can be guaranteed that the obtained solution is also a feasible solution to the chance-constrained problem. Furthermore, this chapter also presents an active methodology based on hypothesis testing using Bayesian inference to identify and disconnect from the agents that perform adversarial actions. In order to decide the connection with the neighbors, each agent must solve a local mixed-integer problem. Note that the probabilistic bounds computed in the passive mechanism are necessary ingredients for the identification scheme. Additionally, how the attack identification and mitigation methods work is shown analytically and by means of simulation.

This chapter is written based on [9–11]. The preliminary idea of both passive and active methods for a restrictive case is presented in [9]. Then Ananduta et al. [10] introduces a stochastic approach and provides the analysis of the methodology for the restrictive case. Finally, Ananduta et al. [11] extend the proposed approach to a more general adversary model. Since the control approach considered in this thesis is based on MPC, it is more related to the work in [1], than that in [2]. However, different from the method proposed in [1], the proposed methodology deals with the attacks by computing control inputs that are robust with respect to such attacks.

Furthermore, the proposed approach is also able to identify the adversarial agents in certain cases and can also deal with more general systems in which there are more than one adversarial agent in the network.

8.2 System and Adversary Models

In this chapter, consider that the subsystems in the network have the capability to operate in the connected and the island modes. This is, in fact, one of the standard features of a microgrid [12]. This feature implies that each subsystem can independently connect or disconnect with its neighbors or the third party. In order to take into account the connections among subsystem in the economic dispatch problem, let $\varrho_{p,t}^q \in \{0, 1\}$ be the connection indicator of agent p with neighbor $q \in \mathcal{N}_{p,t}^{\mathrm{s}}$, i.e., $\varrho_{p,t}^q = 1$ implies agent p connects to neighbor q, whereas $\varrho_{p,t}^q = 0$ implies agent p does not connect to neighbor q. The variable $\varrho_{p,t} = \mathrm{col}(\{\varrho_{p,t}^q\}_{q \in \mathcal{N}_{p,t}^{\mathrm{s}}})$ affects the constraints associated with coupled variables $\boldsymbol{v}_i$, particularly (2.14), which is now redefined, as follows:

$$- v_i^{j,\max} \varrho_{p,t}^q \le v_{i,\tau}^j \le v_i^{j,\max} \varrho_{p,t}^q, \quad \forall j \in \mathcal{N}_i \backslash \mathcal{M}_{p,t}, \ \forall i \in \mathcal{M}_{p,t}, \ \forall \tau \in \mathcal{T}, \tag{8.1}$$

$$- v_i^{j,\max} \le v_{i,\tau}^j \le v_i^{j,\max}, \quad \forall j \in \mathcal{N}_i \cap \mathcal{M}_{p,t}, \ \forall i \in \mathcal{M}_{p,t}, \ \forall \tau \in \mathcal{T}, \tag{8.2}$$

for each subsystem $p \in \mathcal{P}$. Since the preceding constraints define the local set $\mathcal{L}_{i,t}$, the economic dipatch problem (2.15) is restated as follows:

$$\underset{\{\boldsymbol{u}_{i,t}, \boldsymbol{v}_{i,t}\}_{i \in \mathcal{N}}}{\text{minimize}} \ \sum_{p \in \mathcal{P}} \sum_{i \in \mathcal{M}_{p,t}} \left(f_i^{\mathrm{l}}(\boldsymbol{u}_{i,t}) + f_i^{\mathrm{c}}(\boldsymbol{v}_{i,t}) \right) \tag{8.3a}$$

$$\text{s.t. } (\boldsymbol{u}_{i,t}, \boldsymbol{v}_{i,t}) \in \mathcal{L}_{i,t}(\varrho_{p,t}), \ \forall i \in \mathcal{M}_{p,t}, \ \forall p \in \mathcal{P}, \tag{8.3b}$$

$$\boldsymbol{v}_{i,t}^j + \boldsymbol{v}_{j,t}^i = 0, \quad \forall j \in \mathcal{N}_i \cap \mathcal{M}_{p,t}, \ \forall i \in \mathcal{M}_{p,t}, \ \forall p \in \mathcal{P}, \tag{8.3c}$$

$$\boldsymbol{v}_{i,t}^j + \boldsymbol{v}_{j,t}^i = 0, \quad \forall j \in \mathcal{N}_i \backslash \mathcal{M}_{p,t}, \ \forall i \in \mathcal{M}_{p,t}, \ \forall p \in \mathcal{P}. \tag{8.3d}$$

Based on this formulation, the economic dispatch problem considered in the previous chapters considers that each subsystem in the fully connected mode, i.e., $\varrho_{p,t}^q = 1$, for all $q \in \mathcal{N}_{p,t}^{\mathrm{s}}$. Moreover, when agent p is in the island mode, it holds that $\varrho_{p,t}^q = 0$, for all $q \in \mathcal{N}_{p,t}^{\mathrm{s}}$. Later, the capability of disconnecting with neighbors will be used in the identification and mitigation of attacks. Additionally, operating in the island mode requires each subsystem to be self-sufficient, i.e., it can meet its loads using local production for a certain period of time (see Definition 7.2).

Now, denote the decisions at time step t computed by local controllers by $u_{i,t}^{\star}$ and $v_{i,t}^{\star}$, for all $i \in \mathcal{N}$. These decisions are computed by solving Problem (8.3) and several non-centralized algorithms have been presented in the previous chapters. Therefore, assume that one of the non-centralized algorithms is used to compute these decisions. According to the MPC algorithm (step 2 in Algorithm 2.1), these decisions are supposed to be implemented at time step t. However, in this chapter, it is considered that some of the agents might perform adversarial actions. Definition 8.1 classifies the regular and the adversarial agents in the network. In this regard, denote the set of regular agents by $\mathcal{R}$ and the set of adversarial agents by $\mathcal{S}$.

Definition 8.1 An agent $p \in \mathcal{P}$ is regular if it always implements its control inputs $u_{i,t}$ and $v_{i,t}$, for all $i \in \mathcal{M}_{p,t}$, according to the decision computed from the non-centralized algorithm, i.e., $u_{i,t} = u^\star_{i,t}$ and $v_{i,t} = v^\star_{i,t}$, for all $i \in \mathcal{M}_{p,t}$ and $t \geq 0$. Otherwise, agent p is adversarial.

Furthermore, consider the F-local model of adversaries that is stated in Definition 8.2 [4]. Moreover, an attack follows Definition 8.3 and it is assumed that the attacks occur in a probabilistic manner and thus consider Assumption 8.1.

Definition 8.2 Given $F \in \mathbb{Z}_{\geq 1}$, the set of adversarial agents is F-local if $|\mathcal{S} \cap \mathcal{N}^{\mathrm{s}}_{p,t}| \leq F$, for all $p \in \mathcal{P}$. □

Definition 8.3 An attack is an event at which an adversarial agent $p \in \mathcal{S}$ implements a control input that is different than the decision obtained from the non-centralized algorithm, i.e., $u_{i,t} \neq u^\star_{i,t}$ and $v_{i,t} \neq v^\star_{i,t}$, for some $t \in \mathbb{Z}_{\geq 0}$ and $i \in \mathcal{M}_{p,t}$. □

Assumption 8.1 The probability of an attack to occur at each time step $t \in \mathbb{Z}_{\geq 0}$, for each $p \in \mathcal{S}$, can be time-varying with a uniform lower bound, denoted by $\pi^{\mathrm{at}}_p > 0$. □

By performing an attack according to Definition 8.3, an adversarial agent might gain benefit from its neighbors. For instance, an adversarial agent may produce energy with the quantity smaller than the amount that has been decided from the non-centralized algorithm. It then asks its neighbor to compensate the deficiency of power. This attack is possible since these agents are connected and the power balance equations must be met. In this circumstance, the economic cost of the adversarial agent might be less than what it was supposed to be, but its neighbors must pay an extra cost to produce and deliver the energy compensation. It is assumed that if an adversarial agent attacks, it attacks all of its neighbors equally at the same time. Clearly, different distribution schemes can also be considered, but the study of how each adversarial agent attacks is out of the scope of this thesis and is left for future work. The proposed methods presented in Sects. 8.3 and 8.4 can deal with this issue and for this reason this simple assumption is chosen without loss of generality. In the next sections, a non-centralized distributed strategy to deal with such attacks is proposed.

8.3 Robustification Against Attacks

In this section, the passive method, where the computed decisions are robustified against attacks, is described. As discussed in Sect. 2.2, the economic dispatch problem considers non-dispatchable power consumption and production as a source of uncertainty. Similarly, the attacks can also be assumed as another source of uncertainty. Moreover, in Sect. 2.2, it is also assumed that the uncertainty of the loads is bounded and the bound is known. Here, this assumption is relaxed by considering a stochastic approach. To that end, denote by $w_{i,t} = \begin{bmatrix} w_{i,t}^{\mathrm{d}} & w_{i,t}^{\mathrm{at}} \end{bmatrix}^{\top}$ the uncertainties in node $i \in \mathcal{N}$, where $w_{i,t}^{\mathrm{d}}$ is the load uncertainty, as already defined in Sect. 2.2, and $w_{i,t}^{\mathrm{at}}$ is the amount of attack received by node i. Moreover, denote that $\boldsymbol{w}_{i,t} = \mathrm{col}(\{w_{i,\tau}\}_{\tau \in \mathcal{T}})$ and consider Assumption 8.2, which is necessary for a randomization-based method and commonly considered [13, 14]. Note that an assumption on the distribution function of $\boldsymbol{w}_{i,t}$ is not needed, although any type of distribution could be considered as well.

Assumption 8.2 Let $\Omega_i \subseteq \mathbb{R}^{2h}$, for each $i \in \mathcal{N}$, be an uncertain set that is endowed with a Borel σ-algebra. For each node $i \in \mathcal{N}$ and for any $t \in \mathbb{Z}_{\geq 0}$, $\boldsymbol{w}_{i,t} \in \Omega_i$ is a random process that is independent and identically distributed (i.i.d.). □

Furthermore, instead of using (2.5), redefine $\boldsymbol{d}_{i,t}$ by

$$\boldsymbol{d}_{i,t} = \boldsymbol{d}_{i,t}^{\mathrm{f}} + (I_h \otimes \mathbb{1}_2^{\top})\boldsymbol{w}_{i,t}, \quad \forall i \in \mathcal{N}, \tag{8.4}$$

which is used to construct the set $\mathcal{W}_{i,t}(\boldsymbol{w}_{i,t})$, and consequently the local set $\mathcal{L}_{i,t}(\varrho_{p,t}, \boldsymbol{w}_{i,t})$. By taking this into account and since a stochastic method is considered, Problem (8.3) is then reformulated as a chance-constrained problem as follows:

$$\underset{\{\boldsymbol{u}_{i,t}\}_{i \in \mathcal{N}}}{\text{minimize}} \sum_{p \in \mathcal{P}} \sum_{i \in \mathcal{M}_{p,t}} \left(f_i^{\mathrm{l}}(\boldsymbol{u}_{i,t}) + f_i^{\mathrm{c}}(\boldsymbol{v}_{i,t})\right) \tag{8.5a}$$

$$\text{s.t. } \mathbb{P}\big((\boldsymbol{u}_{i,t}, \boldsymbol{v}_{i,t}) \in \mathcal{L}_{i,t}(\varrho_{p,t}, \boldsymbol{w}_{i,t}) | \boldsymbol{w}_{i,t} \in \Omega_i\big) \geq 1 - \varepsilon_i, \ \forall i \in \mathcal{M}_{p,t}, \ \forall p \in \mathcal{P}, \tag{8.5b}$$

$$\boldsymbol{v}_{i,t}^j + \boldsymbol{v}_{j,t}^i = 0, \quad \forall j \in \mathcal{N}_i \cap \mathcal{M}_{p,t}, \ \forall i \in \mathcal{M}_{p,t}, \ \forall p \in \mathcal{P}, \tag{8.5c}$$

$$\boldsymbol{v}_{i,t}^j + \boldsymbol{v}_{j,t}^i = 0, \quad \forall j \in \mathcal{N}_i \backslash \mathcal{M}_{p,t}, \ \forall i \in \mathcal{M}_{p,t}, \ \forall p \in \mathcal{P}. \tag{8.5d}$$

The inequalities in (8.5b) are the chance constraints where $\varepsilon_i \in (0, 1)$ is the maximum allowable level of violation. By having a chance-constrained problem, small probability of violation of the local constraints $(\boldsymbol{u}_{i,t}, \boldsymbol{v}_{i,t}) \in \mathcal{L}_{i,t}(\boldsymbol{w}_{i,t})$ is allowed. Violation, with small probability, in this control level is tolerable since it will only imply suboptimality of the performance. Any solution of Problem (8.5) is referred to as an ε-level feasible solution, where $\varepsilon = \sum_{i \in \mathcal{N}} \varepsilon_i$. To solve Problem (8.5), a two-step stochastic approach [15] is followed. The methodology consists of two steps: first, a probabilistic bound of the disturbance is computed and then, a robust programming problem that takes into account the probabilistic bound is solved. The solutions obtained using this approach are feasible solutions of the corresponding chance-constrained problem with certain confidence [15]. One of the advantages of applying this approach, compared to the standard scenario approach [16], is that since the dimension of the uncertain vector $\boldsymbol{w}_{i,t}$ is smaller than the dimension of the decision vector $(\boldsymbol{u}_{i,t}, \boldsymbol{v}_{i,t})$, the number of scenarios that must be generated is smaller when applying this approach than when directly applying the scenario approach to Problem (8.5) [15]. Furthermore, since the local constraints are convex, the second step of this approach is tractable. The steps of this approach are explained in Sects. 8.3.1 and 8.3.2.

8.3.1 Computing Probabilistic Bounds

In this step, a randomized program, in which a number of scenarios of the disturbance is generated and considered in the constraints of the problem, is solved to compute a set that probabilistically bounds the uncertainty of the chance-constrained problem. Let the set that bounds a portion of the probability mass of $\boldsymbol{w}_{i,t}$ be denoted by $\mathcal{B}^{\star}_{i,t}$, which is defined to be a polyhedral set, i.e., $\mathcal{B}^{\star}_{i,t} = \{\boldsymbol{\psi} \in \mathbb{R}^{2h} : \boldsymbol{\psi}^{\star,\min}_{i,t} \leq \boldsymbol{\psi} \leq \boldsymbol{\psi}^{\star,\max}_{i,t}\}$, where the inequality relations are component-wise, and $\boldsymbol{\psi}^{\star,\min}_{i,t}, \boldsymbol{\psi}^{\star,\max}_{i,t} \in \mathbb{R}^{2h}$ denote the lower and upper bounds of $\boldsymbol{w}_{i,t}$, respectively. Moreover, note that $\boldsymbol{\psi}^{\star,\min}_{i,t} = \mathrm{col}(\{\psi^{\star,\min}_{i,\tau}\}_{\tau\in\mathcal{T}})$, where $\psi^{\star,\min}_{i,\tau} = \left[\psi^{\mathrm{d}\star,\min}_{i,\tau} \; \psi^{\mathrm{at}\star,\min}_{i,\tau}\right]^{\top}$, and similarly, $\boldsymbol{\psi}^{\star,\max}_{i,t} = \mathrm{col}(\{\psi^{\star,\max}_{i,\tau}\}_{\tau\in\mathcal{T}})$, where $\psi^{\star,\max}_{i,\tau} = \left[\psi^{\mathrm{d}\star,\max}_{i,\tau} \; \psi^{\mathrm{at}\star,\max}_{i,\tau}\right]^{\top}$, where $\psi^{\mathrm{d}\star,\min}_{i,\tau}$ and $\psi^{\mathrm{d}\star,\max}_{i,\tau}$ are associated with $w^{\mathrm{d}}_{i,\tau}$ whereas $\psi^{\mathrm{at}\star,\min}_{i,\tau}$ and $\psi^{\mathrm{at}\star,\max}_{i,\tau}$ are associated with $w^{\mathrm{at}}_{i,\tau}$.

In order to compute $\mathcal{B}^{\star}_{i,t}$, a chance-constrained problem for each $i \in \mathcal{N}$ is formulated as follows:

$$\begin{aligned} \underset{\boldsymbol{\psi}^{\min}_{i,t},\boldsymbol{\psi}^{\max}_{i,t}}{\text{minimize}} \;& \mathbb{1}^{\top}_{2h}\left(\boldsymbol{\psi}^{\max}_{i,t} - \boldsymbol{\psi}^{\min}_{i,t}\right) \\ \text{s.t.}\;& \mathbb{P}\left(\boldsymbol{w}_{i,t} \in [\boldsymbol{\psi}^{\min}_{i,t}, \boldsymbol{\psi}^{\max}_{i,t}] | \boldsymbol{w}_{i,t} \in \Omega_i\right) \geq 1 - \varepsilon_i. \end{aligned} \tag{8.6}$$

Problem (8.6) is solved with the scenario approach, which can be stated as follows:

$$\begin{aligned} \underset{\boldsymbol{\psi}^{\min}_{i,t},\boldsymbol{\psi}^{\max}_{i,t}}{\text{minimize}} \;& \mathbb{1}^{\top}_{2h}\left(\boldsymbol{\psi}^{\max}_{i,t} - \boldsymbol{\psi}^{\min}_{i,t}\right) \\ \text{s.t.}\;& \boldsymbol{w}^{(s)}_{i,t} \in [\boldsymbol{\psi}^{\min}_{i,t}, \boldsymbol{\psi}^{\max}_{i,t}], \quad s = 1, \ldots, n_{s,i}, \end{aligned} \tag{8.7}$$

where $\boldsymbol{w}^{(s)}_{i,t}$ denote a scenario of $\boldsymbol{w}_{i,t}$, generated according to the probability measure for $\boldsymbol{w}_{i,t}$ in an i.i.d. manner, and $n_{s,i}$ is the number of scenarios, which satisfies [15]

$$n_{s,i} \geq \frac{e}{\varepsilon_i(e-1)}\left(4h - 1 + \ln\frac{1}{\zeta_i}\right), \tag{8.8}$$

where $\zeta_i \in (0, 1)$ indicates the desired level of confidence and e is the Euler constant. With some abuse of notation, let the set $\mathcal{B}^{\star}_{i,t}$ be constructed from the solution of Problem (8.7). This set is a feasible solution of (8.6) with probability at least $1 - \zeta_i$.

Remark 8.1 The level of violation (ε_i) and the level of confidence (ζ_i) are predefined parameters that determine the number of scenarios needed to be generated. □

8.3.2 Robust Reformulation

Upon obtaining the bounds $\mathcal{B}^{\star}_{i,t}$, for all $i \in \mathcal{N}$, a robust counterpart of Problem (8.5) is derived. Since the local constraints are convex, the vertex enumeration method, where the uncertain variable $\boldsymbol{w}_{i,t}$, for each $i \in \mathcal{N}$, is substituted with the vertices of $\mathcal{B}^{\star}_{i,t}$, can be applied [15]. Particularly, the worst-case disturbance based on the computed bounds $\mathcal{B}^{\star}_{i,t}$ is $\boldsymbol{\psi}^{\star,\max}_{i,t}$. In other words, $\boldsymbol{\psi}^{\star,\max}_{i,t}$ becomes a probabilistic upper-bound of $\boldsymbol{w}_{i,t}$. Therefore, the robust formulation associated with Problem (8.5) is stated as follows:

$$\underset{\{(\boldsymbol{u}_{i,t},\boldsymbol{v}_{i,t})\}_{i\in\mathcal{N}}}{\text{minimize}} \sum_{p\in\mathcal{P}} \sum_{i\in\mathcal{M}_{p,t}} \left(f_i^{\mathrm{l}}(\boldsymbol{u}_{i,t}) + f_i^{\mathrm{c}}(\boldsymbol{v}_{i,t})\right) \tag{8.9a}$$

$$\text{s.t. } (\boldsymbol{u}_{i,t}, \boldsymbol{v}_{i,t}) \in \mathcal{L}_{i,t}(\varrho_{p,t}, \boldsymbol{\psi}^{\star,\max}_{i,t}), \quad \forall i \in \mathcal{M}_{p,t},\ \forall p \in \mathcal{P}, \tag{8.9b}$$

$$\boldsymbol{v}^{j}_{i,t} + \boldsymbol{v}^{i}_{j,t} = 0, \quad \forall j \in \mathcal{N}_i \cap \mathcal{M}_{p,t},\ \forall i \in \mathcal{M}_{p,t},\ \forall p \in \mathcal{P}, \tag{8.9c}$$

$$\boldsymbol{v}^{j}_{i,t} + \boldsymbol{v}^{i}_{j,t} = 0, \quad \forall j \in \mathcal{N}_i \backslash \mathcal{M}_{p,t},\ \forall i \in \mathcal{M}_{p,t},\ \forall p \in \mathcal{P}. \tag{8.9d}$$

In regard to Problem (8.9) and the island mode operation of the agents, consider that Assumption 8.3, which replaces Assumption 2.3, holds.

Assumption 8.3 The feasible set of Problem (8.9) is non-empty. Furthermore, this set has a subset in which $\boldsymbol{v}^{j}_{i,t} = \boldsymbol{v}^{i}_{j,t} = 0$, for all $j \in \mathcal{N} \backslash \mathcal{M}_p$, $i \in \mathcal{M}_p$, and $p \in \mathcal{P}$.

The existence of nonempty feasible region in Problem (8.9) depends on the scenario realization, which determines the construction of the probabilistic bounds. Therefore, each agent can compute the bounds such that Assumption 8.3 holds. Furthermore, the second part of Assumption 8.3 follows from the consideration that each agent can operate in the island mode, which requires self-sufficiency. The robustness of the decisions obtained by solving (8.9) is characterized by [15, Proposition 1], as follows.

Proposition 8.1 *Suppose that Assumptions 8.2 and 8.3 hold. Furthermore, let $\boldsymbol{\psi}^{\star,\max}_{i,t}$, for all $i \in \mathcal{N}$, are computed by solving (8.7), where the number of generated scenarios satisfies (8.8). Then, the solution to Problem (8.9) is an ε-level feasible solution of Problem (8.5), with probability at least $1 - \zeta$, where $\varepsilon = \sum_{i\in\mathcal{N}} \varepsilon_i$ and $\zeta = \sum_{i\in\mathcal{N}} \zeta_i$.* □

Therefore, Problem (8.9) is solved to compute $\boldsymbol{u}_{i,t}^{\star}$ and $\boldsymbol{v}_{i,t}^{\star}$, for all $i \in \mathcal{N}$, and the non-centralized methods that are proposed in Chaps. 3, 5, and 7 can be used. Note that when the non-centralized scheme in Chap. 7 is used, since each agent is self-sufficient due to the second part of Assumption 8.3, the network is not repartitioned and $\mathcal{N}_{p,t}^{\mathrm{s}}$ for each $p \in \mathcal{P}$, is constant, whereas the other distributed methods do not change the partitions. Therefore, for the remaining of the chapter, it is assumed that the set of neighbors of each agent $p \in \mathcal{P}$ is fixed and denoted by $\mathcal{N}_p^{\mathrm{s}}$.

8.4 Attack Identification and Mitigation Method

Besides employing the stochastic approach outlined in Sect. 8.3, an active methodology to identify the adversarial agents and mitigate the attacks is also proposed. The identification methodology is a hypothesis testing using Bayesian inference. Furthermore, the identification method requires regular agents to actively disconnect the links with their neighbors. To decide from which neighbors an agent should disconnect, a local mixed-integer optimization problem is solved.

8.4.1 Attack Detection

In order to identify the adversarial neighbors, a regular agent must be able to detect an attack in the first place. Now, assume that an agent can measure the total disturbance at each step t. This assumption is justifiable since each agent $p \in \mathcal{P}$ has the information of the set points $\boldsymbol{u}_{i,t}^{\star}$ and $\boldsymbol{v}_{i,t}^{\star}$, for all $i \in \mathcal{M}_{p,t}$, and can obtain the actual energy consumption at time step t, by measurements. Thus, the difference between them, which can be regarded as power imbalance, shows the actual total disturbance affecting the agent. Denote the total disturbance of agent p by $\omega_{p,t}$, then it holds that

$$\omega_{p,t} = \sum_{i \in \mathcal{M}_p} \left(\mathbb{1}^\top \boldsymbol{u}_{i,t}^{\star} + \mathbb{1}^\top \boldsymbol{v}_{i,t}^{\star} - d_{i,t}^{\mathrm{ac}} \right), \tag{8.10}$$

where $d_{i,t}^{\mathrm{ac}}$ is the actual load of node $i \in \mathcal{M}_{p,t}$ measured at time t. As can be seen, (8.10) actually evaluates the power balance of all the nodes that belong to agent p. Note that, as mentioned in Sect. 2.3, the power mismatches at each node is handled by low-level controllers, which act similarly as an automatic generation control [14, 17].

The probabilistic bound $\mathcal{B}_{i,t}^{\star}$ can now be used as the threshold to define whether an attack occurs. Recall that $\psi_{i,t}^{\star,\min} = [\psi_{i,t}^{\mathrm{d}\star,\min}\ \psi_{i,t}^{\mathrm{at}\star,\min}]^\top$, where $\psi_{i,t}^{\mathrm{d}\star,\min}$ and $\psi_{i,t}^{\mathrm{at}\star,\min}$ denote the lower bounds of the disturbance associated to the load and to an attack, respectively, and similarly $\boldsymbol{\psi}_{i,t}^{\star,\max} = [\psi_{i,t}^{\mathrm{d}\star,\max}\ \psi_{i,t}^{\mathrm{at}\star,\max}]^\top$, where $\psi_{i,t}^{\mathrm{d}\star,\max}$ and $\psi_{i,t}^{\mathrm{at}\star,\max}$ denote the corresponding upper bounds. Then, the attack detection is defined as follows.

Definition 8.4 Let $\delta^{\text{at}}_{p,t} \in \{0, 1\}$, for each regular agent $p \in \mathcal{R}$ and $t \in \mathbb{Z}_{\geq 0}$, be the indicator that detects attacks. If

$$\sum_{i \in \mathcal{M}_{p,t}} \psi^{\text{d}\star,\min}_{i,t} \leq \omega_{p,t} \leq \sum_{i \in \mathcal{M}_{p,t}} \psi^{\text{d}\star,\max}_{i,t}, \tag{8.11}$$

then there is no attack detected and $\delta^{\text{at}}_{p,t} = 0$. Otherwise, $\delta^{\text{at}}_{p,t} = 1$, implying an attack is detected. □

Using Definition 8.4, the following definitions of an undetectable attack and a false attack detection are given. Furthermore, Assumption 8.4 related to undetectable attacks is considered.

Definition 8.5 An attack is detectable if $\sum_{i \in \mathcal{M}_{p,t}} w^{\text{at}}_{i,t} \neq 0$ such that (8.11) does not hold. On the other hand, an attack is undetectable if $\sum_{i \in \mathcal{M}_{p,t}} w^{\text{at}}_{i,t} \neq 0$ such that (8.11) holds. □

Assumption 8.4 The probability of the undetectable attacks that are received by agent $p \in \mathcal{R}$, denoted by π^{ua}_p, is less than 1. □

Definition 8.6 A false attack detection occurs at agent p when $w^{\text{at}}_{i,t} = 0$, for all $i \in \mathcal{M}_{p,t}$ and $\delta^{\text{at}}_{p,t} = 1$, i.e., if $\sum_{i \in \mathcal{M}_{p,t}} w^{\text{d}}_{i,t} < \sum_{i \in \mathcal{M}_{p,t}} \psi^{\text{d}\star,\min}_{i,t}$ or $\sum_{i \in \mathcal{M}_{p,t}} w^{\text{d}}_{i,t} > \sum_{i \in \mathcal{M}_{p,t}} \psi^{\text{d}\star,\max}_{i,t}$ when $w^{\text{at}}_{i,t} = 0$, for all $i \in \mathcal{M}_{p,t}$. □

When an attack is undetectable, the regular agent $p \in \mathcal{R}$ cannot distinguish the attack from the load disturbance. Such an attack is tolerable since the *total* disturbance is within the bounds of the *load* disturbance. For systems without disturbances, undetectable attacks are also defined similarly in [3]. Furthermore, Assumption 8.4 implies that some attacks are detectable by agent p, which is necessary in order to identify the adversarial neighbor. On the other hand, a false detection occurs when the load disturbance is outside of the computed bounds. It is worth mentioning that the detection scheme is similar to the passive fault detection method presented in [18].

8.4.2 *Identification and Mitigation Method*

A regular agent that has more than one neighbor is not able to identify which ones are adversarial although it could detect an attack based on Definition 8.4. Therefore, a hypothesis testing scheme based on Bayesian inference is formulated as the method to identify the adversarial neighbors. In this method, it is assumed that regular agents do not have prior knowledge of the occurrence of the attacks, but each agent $p \in \mathcal{R}$ has an initial expectation on the probability of attacks received, denoted by $\hat{\pi}^{\text{at}}_p$. The parameter $\hat{\pi}^{\text{at}}_p$ is a positive constant smaller than one that is used to update the hypothesis probability as shown in (22)–(23) below. It does not need to reflect the actual probability of the received attacks and only affects the convergence of the identification process as shown in Sect. 8.5.

Based on the F-local adversary model, where $|\mathcal{N}_p^{\mathrm{s}}| \geq F$, each regular agent $p \in \mathcal{R}$ considers all the sets that are the elements of the power set of $\mathcal{N}_p^{\mathrm{s}}$ and have the cardinality F. Let these sets be denoted by Θ_p^{χ}, for $\chi = 1, 2, \ldots, n_p^{\mathrm{H}}$, where $n_p^{\mathrm{H}} = \binom{|\mathcal{N}_p^{\mathrm{s}}|}{F}$. Note that $\Theta_p^{\chi_1} \neq \Theta_p^{\chi_2}$ if $\chi_1 \neq \chi_2$. Now, consider a set of hypotheses, $\mathcal{H}_p = \{\mathbf{H}_p^{\chi} : \chi = 0, 1, \ldots, n_p^{\mathrm{H}}\}$, where the hypotheses are defined as follows:

- $\mathbf{H}_p^0$: There is no adversarial neighbor,
- $\mathbf{H}_p^{\chi}$: The set Θ_p^{χ} contains all the adversarial neighbors,

for $\chi = 1, \ldots, n_p^{\mathrm{H}}$. Hence, each set Θ_p^{χ} is associated with one hypothesis. Furthermore, recall the attack indicator variable $\delta_{p,t}^{\mathrm{at}}$, which is defined in Definition 8.4. The Bayesian inference is used as the model to update the probability of the hypothesis as follows:

$$\mathbb{P}_{t+1}(\mathbf{H}_p^{\chi}) = \frac{\mathbb{P}_t(\mathbf{H}_p^{\chi})\mathbb{P}_t(\delta_{p,t}^{\mathrm{at}}|\mathbf{H}_p^{\chi})}{\mathbb{P}_t(\delta_{p,t}^{\mathrm{at}})}, \quad \forall \mathbf{H}_p^{\chi} \in \mathcal{H}_p, \tag{8.12}$$

where $\mathbb{P}_{t+1}(\mathbf{H}_p^{\chi})$ is the a posteriori probability of $\mathbf{H}_p^{\chi}$ given the event $\delta_{p,t}^{\mathrm{at}}$, i.e., $\mathbb{P}_{t+1}(\mathbf{H}_p^{\chi}) = \mathbb{P}(\mathbf{H}_p^{\chi}|\delta_{p,t}^{\mathrm{at}})$; $\mathbb{P}_t(\mathbf{H}_p^{\chi})$ denotes the probability of hypothesis $\mathbf{H}_p^{\chi}$ at time instant t; $\mathbb{P}_t(\delta_{p,t}^{\mathrm{at}})$ denotes the marginal likelihood of $\delta_{p,t}^{\mathrm{at}}$; and $\mathbb{P}_t(\delta_{p,t}^{\mathrm{at}}|\mathbf{H}_p^{\chi})$ denotes the probability of observing $\delta_{p,t}^{\mathrm{at}}$ given hypothesis $\mathbf{H}_p^{\chi}$. The probability $\mathbb{P}_t(\delta_{p,t}^{\mathrm{at}}|\mathbf{H}_p^{\chi})$ is formulated as follows:

$$\begin{aligned}
\mathbb{P}_t(\delta_{p,t}^{\mathrm{at}} = 0|\mathbf{H}_p^{\chi}) &= \begin{cases} 1, & \text{for } \chi = 0, \\ 1 - \left(\max_{q \in \Theta_p^{\chi}} \varrho_{p,t}^{q}\right) \hat{\pi}_p^{\mathrm{at}}, & \text{for } \chi = 1, \ldots, n_p^{\mathrm{H}}, \end{cases} \\
\mathbb{P}_t(\delta_{p,t}^{\mathrm{at}} = 1|\mathbf{H}_p^{\chi}) &= \begin{cases} 0, & \text{for } m = 0, \\ \left(\max_{q \in \Theta_p^{\chi}} \varrho_{p,t}^{q}\right) \hat{\pi}_p^{\mathrm{at}}, & \text{for } \chi = 1, \ldots, n_p^{\mathrm{H}}, \end{cases}
\end{aligned} \tag{8.13}$$

where $\varrho_{p,t}^{q} \in \{0, 1\}$, for all $q \in \mathcal{N}_p^{\mathrm{s}}$, denote the decision whether agent p both connects to and negotiates with neighbor q. Additionally, the initial probabilities of all hypotheses are defined as

$$\mathbb{P}_0(\mathbf{H}_p^{\chi}) = \begin{cases} 1 - \hat{\pi}_p^{\mathrm{at}}, & \text{for } \chi = 0, \\ \hat{\pi}_p^{\mathrm{at}}/n_p^{\mathrm{H}} & \text{for } \chi = 1, \ldots, n_p^{\mathrm{H}}, \end{cases} \tag{8.14}$$

implying that agent p initially considers each neighbor has an equal chance of attacking. Note that $\hat{\pi}_p^{\mathrm{at}}$ does not need to be equal to any p_q^{at}, for $q \in \mathcal{N}_p^{\mathrm{s}} \cap \mathcal{S}$. In addition, for agent $p \in \mathcal{R}$, where $|\mathcal{N}_p^{\mathrm{s}}| < F$, two hypotheses are considered, i.e., $\mathbf{H}_p^0$ and $\mathbf{H}_p^1$, where $\Theta_p^1 = \mathcal{N}_p^{\mathrm{s}}$.

The last ingredient of the active method is the connection decision at each step, $\varrho_{p,t}$. In order to compute $\varrho_{p,t}$, agent p solves a local mixed-integer optimization problem as follows:

$$\underset{\varrho_p, \{(\boldsymbol{u}_{i,t}, \boldsymbol{v}_{i,t})\}_{i \in \mathcal{M}_{p,t}}}{\text{minimize}} \quad \gamma_p n^{\text{at}}_{p,t} \sum_{q \in \mathcal{N}^{\text{s}}_p} c^q_{p,t} \varrho^q_p + \sum_{i \in \mathcal{M}_{p,t}} f^{\text{l}}_i(\boldsymbol{u}_{i,t}) \tag{8.15a}$$

$$\text{s.t.} \quad (\boldsymbol{u}_{i,t}, \boldsymbol{v}_{i,t}) \in \mathcal{L}_{i,t}(\varrho_p, \boldsymbol{\psi}^{\star,\max}_{i,t}), \quad \forall i \in \mathcal{M}_{p,t}, \ \forall p \in \mathcal{P}, \tag{8.15b}$$

$$\boldsymbol{v}^j_{i,t} + \boldsymbol{v}^i_{j,t} = 0, \quad \forall j \in \mathcal{N}_i \cap \mathcal{M}_{p,t}, \ \forall i \in \mathcal{M}_{p,t}, \tag{8.15c}$$

$$\varrho_p \in \{\mathbb{1}_{|\mathcal{N}^{\text{s}}_p|}\} \cup \{\varrho \in \{0,1\}^{|\mathcal{N}^{\text{s}}_p|} : \mathbb{1}^\top \varrho = |\mathcal{N}^{\text{s}}_p| - F\}, \tag{8.15d}$$

where $\varrho_p = \text{col}(\{\varrho^q_p\}_{q \in \mathcal{N}^{\text{s}}_p})$ and $\varrho^q_p \in \{0,1\}$. Furthermore, $c^q_{p,t} = \sum_{\chi_q \in \{\chi : q \in \Theta^{\chi}_p\}} \mathbb{P}_t(\mathbf{H}^{\chi_q}_p)$, for each $q \in \mathcal{N}^{\text{s}}_p$, denotes the individual weight associated to ϱ^q_p, $\gamma_p \in \mathbb{R}_{>0}$ denotes a weight that must be predefined and $n^{\text{at}}_{p,t}$ denotes the number of attacks that agent p has detected, i.e., $n^{\text{at}}_{p,t} = \sum_{\tau=0}^{t} \delta^{\text{at}}_{p,\tau}$. By having $n^{\text{at}}_{p,t}$ as a weight, establishing a connection with a neighbor is penalized more if the number of attacks detected increases. Moreover, the constraint (8.15d) implies that agent p can either connect to all neighbors or disconnect from any F neighbors. It can be seen that Problem (8.15) considers the same local constraints of Problem (8.9) and disregards the coupling constraints among subsystems.

Problem (8.15) is a mixed-integer quadratic program (MIQP). Notice that the Boolean variable $\varrho^q_{p,t}$, for each $q \in \mathcal{N}^{\text{s}}_p$, is penalized by adding weight $c^q_{p,t}$, proportionally to the sum of probability measures of the hypotheses associated with neighbor q. Furthermore, having constraint (8.15d) means that there are only $n^{\text{H}}_p + 1$ possible solutions of ϱ_p. Therefore, if n^{H}_p is relatively small, agent p might solve $n^{\text{H}}_p + 1$ convex problems, i.e., Problem (8.15) without (8.15d) and with a fixed and different ϱ_p that satisfies (8.15d). Another way to solve Problem (8.15) is by directly employing a mixed-integer optimization method such as the branch-and-bound algorithm. Finally, let $\mathcal{Q}^\star_{p,t} \in \{0,1\}^{|\mathcal{N}^{\text{s}}_p|}$ be the set of minimizers of Problem (8.15) and suppose that the decision $\varrho_{p,t}$ is chosen from $\mathcal{Q}^\star_{p,t}$, i.e., $\varrho_{p,t} \in \mathcal{Q}^\star_{p,t}$.

Remark 8.2 One might set γ_p large enough such that $\mathbb{1}_{|\mathcal{N}^{\text{s}}_p|} \in \mathcal{Q}^\star_{p,t}$ only if $n^{\text{at}}_{p,t} = 0$. In this case, once an attack is detected, some of the neighbors are always disconnected. □

Remark 8.3 Based on the second part of Assumption 8.3, the feasibility of Problem (8.15) is guaranteed.

8.5 The Overall Scheme

The inclusion of the proposed methodologies in this chapter to the non-centralized MPC scheme (Algorithm 2.1) is summarized in Algorithm 8.1. First, each regular agent $P \in \mathcal{R}$ computes the probabilistic bounds of the disturbance and decides the connections with the neighbors. Then, it performs a non-centralized algorithm, e.g., the DAL method (Algorithm 3.1) or the distributed ADMM-based method (Algorithm 3.2) to compute the decisions. After the decisions are applied, the agent detects the occurrence of an attack and performs the identification method. Note that $\mathbb{P}(\mathbf{H}_p^{\hat{\chi}})$ in step 4 of Algorithm 8.1 is associated to the set $\Theta_p^{\hat{\chi}} = \{q \in \mathcal{N}_p^{\mathrm{s}} : \varrho_{p,t}^q = 0\}$. Related to step 6 of Algorithm 8.1, in particular the part of implementing $\varrho_{p,t}$, it has been assumed that any agent can temporarily disconnect the physical link between itself and its neighbors, respecting the decision of $\varrho_{p,t}$. Note that two agents, p and q, where $\{p, q\} \in \mathcal{E}^{\mathrm{s}}$, can only exchange energy if and only if $\varrho_{p,t}^q = \varrho_{q,t}^p = 1$. Therefore, although there exists a connection between agents p and q, either of them can block the influence by closing the connection. Furthermore, due to constraints (8.1), when one of two neighboring agents, say p and q, decides to disconnect, the negotiation of the power transferred is forced towards a common solution $\boldsymbol{v}_{i,t}^j = \boldsymbol{v}_{j,t}^i = 0$, for any $i \in \mathcal{M}_{p,t}$ and $j \in \mathcal{M}_{q,t}$ where $\{i, j\} \in \mathcal{E}$. Although this assumption is not suitable for a conventional power network, considering the framework of microgrids, which can work in the island mode, disconnecting two neighboring microgrids can be done. The analysis of the algorithm, in terms of the result of the mitigation and identification method, is presented next in Sect. 8.6.

Remark 8.4 In the problem setting, for simplicity, the probability distribution of the uncertainties $\boldsymbol{w}_{i,t}$ remains the same over time under Assumption 2. In this case, to reduce the computational effort at each iteration in Algorithm 8.1, steps 3 and 4 can actually be carried out only once offline [13]. In this case, suppose that the probabilistic bounds that are computed by solving (8.7) in an offline manner are denoted

---

Algorithm 8.1 Resilient Non-centralized MPC, for $p \in \mathcal{R}$

Initialize the hypothesis probabilities according to (8.14).
Iteration: for $t = 0, 1, 2, \ldots$

1. Choose $n_{s,i}$ according to (8.8).
2. Compute the probabilistic bound $\mathcal{B}_{i,t}^{\star}$, for all $i \in \mathcal{M}_{p,t}$, by solving (8.7).
3. Compute $\mathcal{Q}_{p,t}^{\star}$ by solving (8.15).
4. If $\mathbb{1}_{|\mathcal{N}_p^{\mathrm{s}}|} \in \mathcal{Q}_{p,t}^{\star}$, then choose $\varrho_{p,t} = \mathbb{1}_{|\mathcal{N}_p^{\mathrm{s}}|}$. Otherwise, choose randomly $\varrho_{p,t} \in \mathcal{Q}_{p,t}^{\star}$ such that $\mathbb{P}(\mathbf{H}_p^{\hat{\chi}}) \neq 0$.
5. Compute $u_{i,t}^{\star}$ and $v_{i,t}^{\star}$, for all $i \in \mathcal{M}_{p,t}$, by solving (8.9) using a non-centralized algorithm.
6. Implement $u_{i,t}^{\star}$ and $v_{i,t}^{\star}$, for all $i \in \mathcal{M}_{p,t}$, and $\varrho_{p,t}$.
7. Measure the states, $x_{i,t+1}$, for all $i \in \mathcal{M}_{p,t} \cap \mathcal{N}^{\mathrm{st}}$.
8. Compute $\delta_{p,t}^{\mathrm{at}}$ based on Definition 8.4.
9. Update the probability values of the hypotheses according to (8.12).

---

by $\mathcal{B}_i^\star$, for all $i \in \mathcal{N}$. Then, at each time instant t, $\mathcal{B}_{i,t}^\star = \mathcal{B}_i^\star$. In general, however, the distribution of $\boldsymbol{w}_{i,t}$ can be time-varying, and with sufficient computational resources and knowledge regarding the variables, steps 3 and 4 can be executed online.

8.6 Analysis of the Identification Method

This section shows how each regular agent can correctly identify the adversarial agent in the case $F = 1$ and, in general, for any F-local adversary model, can block the influence of all adversarial neighbors by employing Algorithm 8.1. Note that in this model, the total number of adversarial agents in the network might be more than F. The analysis is divided into two parts, for the case $F = 1$ and for the case $F > 1$. Firstly, the following lemmas, which are useful in the analysis, are established.

Lemma 8.1 *Suppose that Assumptions 8.1–8.4 hold. If a regular agent, $p \in \mathcal{R}$, is connected to any adversarial neighbor, then the probability that infinitely many detectable attacks are received by agent p is 1.* □

Proof Since Assumptions 8.1 and 8.4 hold, the probability that detectable attacks occur is a positive scalar that is lower bounded by $\min_{q \in \mathcal{N}_p^{\mathrm{s}} \cap \mathcal{S}} \pi_q^{\mathrm{at}}(1 - \pi_p^{\mathrm{ua}}) > 0$. Furthermore, the connection between agent p and an adversarial neighbor implies that agent q can receive an attack from the adversarial neighbor. As a result, based on the Borel-Cantelli lemma [19, Section 8.2.3] and since Assumption 8.2 holds, the claimed statement follows. □

Lemma 8.2 *Let Assumption 8.3 hold. Suppose that at time step τ, the minimizer $\boldsymbol{v}_{i,\tau}^\star$ chosen from the set of minimizers $\mathcal{Q}_{i,\tau}^\star$ of Problem (8.15) is such that some of the neighbors are disconnected, i.e., there exists some $q^\star \in \Theta_p^{\hat{\chi}} = \{q \in \mathcal{N}_p^{\mathrm{s}} : \varrho_{p,t}^q = 0\}$, $|\Theta_p^{\hat{\chi}}| = F$. Furthermore, suppose that for $t \geq \tau$, $\mathbb{1}_{|\mathcal{N}_p^{\mathrm{s}}|} \notin \mathcal{Q}_{p,t}^\star$. If agent p does not detect an attack, i.e., $\delta_{p,\tau}^{\mathrm{at}} = 0$, then $\mathcal{Q}_{i,\tau+1}^\star = \{\varrho_{p,\tau}\}$.*

Proof The decision about which neighbors should be disconnected by agent p is based on the weight of $\varrho_{p,t}^q$ in (8.15a). For the case $F = 1$, the weight depends on the probability measure of each hypothesis, i.e., $c_{p,\tau}^q = \mathbb{P}_\tau(\mathbf{H}_p^{\chi_q})$, where $\mathbf{H}_p^{\chi_q}$, for each $q \in \mathcal{N}_p^{\mathrm{s}}$, is associated to $\Theta_p^{\chi_q} = \{q\}$. Therefore, $\varrho_{p,\tau}^{q^\star} = 0$ implies that $\mathbb{P}_\tau(\mathbf{H}_p^{\chi_{q^\star}}) \in \arg\max_{q \in \mathcal{N}_p^{\mathrm{s}}}(\mathbb{P}_\tau(\mathbf{H}_p^{\chi_q}))$. Observe that, for $t \geq \tau$, $\mathbb{P}_t(\delta_{p,t}^{\mathrm{at}} = 0) = 1$ only if $\mathbb{P}_\tau(\mathbf{H}_p^{\chi_{q^\star}}) = 1$, implying $\mathbb{P}_\tau(\mathbf{H}_p^{\chi_q}) = 0$ for $q \neq q^\star$. Thus in this case the claimed statement follows immediately. Now, consider the case where $\mathbb{P}_\tau(\mathbf{H}_p^{\chi_{q^\star}}) < 1$. Since $d_{i,\tau}^{\mathrm{at}} = 0$ and $\mathbb{P}_t(\delta_{p,t}^{\mathrm{at}}) \in (0, 1)$, by applying (8.12), it follows that $\mathbb{P}_{\tau+1}(\mathbf{H}_p^{\chi_{q^\star}}) > \mathbb{P}_\tau(\mathbf{H}_p^{\chi_{q^\star}})$ and $\mathbb{P}_{\tau+1}(\mathbf{H}_p^{\chi_q}) \leq \mathbb{P}_\tau(\mathbf{H}_p^{\chi_q}) \leq \mathbb{P}_\tau(\mathbf{H}_p^{\chi_{q^\star}})$, for any $q \in \mathcal{N}_p^{\mathrm{s}} \backslash \{q^\star\}$. Therefore, $\mathbb{P}_{\tau+1}(\mathbf{H}_p^{\chi_{q^\star}}) = \arg\max_{q \in \mathcal{N}_p^{\mathrm{s}}}(\mathbb{P}_{\tau+1}(\mathbf{H}_p^{\chi_q}))$ and the claimed statement follows.

For the case $F > 1$, let $\Theta_{p,\tau}^{\hat{\chi}}$ be the set of all neighbors that are disconnected from agent p at time step $t = \tau$, i.e., $\Theta_{p,\tau}^{\hat{\chi}} = \{q \in \mathcal{N}_p^{\mathrm{s}} : \varrho_{p,\tau}^q = 0\}$. Note that there is a hypothesis that is associated to $\Theta_{p,\tau}^{\hat{\chi}}$, denoted by $\mathbf{H}_p^{\hat{\chi}}$, and the probability measure of this hypothesis is denoted by $\mathbb{P}_t(\mathbf{H}_p^{\hat{\chi}})$. Consider any $q^\star \in \Theta_{p,\tau}^{\hat{\chi}}$ and $q \in \mathcal{N}_p^{\mathrm{s}} \backslash \Theta_{p,\tau}^{\hat{\chi}}$. Since $q^\star \in \Theta_{p,\tau}^{\hat{\chi}}$ are disconnected at $t = \tau$, it holds that $c_{p,\tau}^{q^\star} \geq c_{p,\tau}^{q}$, for any $q \in \mathcal{N}_{p,t}^{\mathrm{s}}$. Now, it will be shown that $c_{p,\tau+1}^{q^\star} > c_{p,\tau+1}^{q}$ by updating $\mathbb{P}_{\tau+1}(\mathbf{H}_p^{\chi})$ for all $\mathbf{H}_p^{\chi} \in \mathcal{H}_p$ with (8.12) when $\delta_{p,\tau}^{\mathrm{at}} = 0$. Similarly to the case $F = 1$, if for any $q \in \mathcal{N}_p^{\mathrm{s}} \backslash \Theta_{p,\tau}^{\hat{\chi}}$, $c_{p,\tau}^{q} = 0$, it holds that $c_{p,\tau+1}^{q^\star} > c_{p,\tau+1}^{q} = c_{p,\tau}^{q} = 0$. In the case that $c_{p,\tau}^{q} > 0$, for some $q \in \mathcal{N}_p^{\mathrm{s}} \backslash \Theta_{p,\tau}^{\hat{\chi}}$, it follows that $c_{p,\tau+1}^{q^\star} > c_{p,\tau+1}^{q}$, which is shown as follows.

From the fact that $c_{p,\tau}^{q^\star} \geq c_{p,\tau}^{q}$, it holds that

$$\begin{aligned} c_{p,\tau}^{q^\star} \geq c_{p,\tau}^{q} &\Leftrightarrow \mathbb{P}_\tau(\mathbf{H}_p^{\hat{\chi}}) + \sum_{\chi \in \mathcal{J} \backslash \{\hat{\chi}\}} \mathbb{P}_\tau(\mathbf{H}_p^{\chi}) + \sum_{\chi \in \hat{\mathcal{J}}} \mathbb{P}_\tau(\mathbf{H}_p^{\chi}) \geq \sum_{\chi \in \mathcal{J}'} \mathbb{P}_\tau(\mathbf{H}_p^{\chi}) + \sum_{\chi \in \hat{\mathcal{J}}} \mathbb{P}_\tau(\mathbf{H}_p^{\chi}) \\ &\Leftrightarrow \mathbb{P}_\tau(\mathbf{H}_p^{\hat{\chi}}) + \sum_{\chi \in \mathcal{J} \backslash \{\hat{\chi}\}} \mathbb{P}_\tau(\mathbf{H}_p^{\chi}) - \sum_{\chi \in \mathcal{J}'} \mathbb{P}_\tau(\mathbf{H}_p^{\chi}) \geq 0, \end{aligned} \tag{8.16}$$

where $\mathcal{J} = \{\chi : q^\star \in \Theta_p^{\chi}, q \notin \Theta_p^{\chi}\}$, $\mathcal{J}' = \{\chi : q \in \Theta_p^{\chi}, q^\star \notin \Theta_p^{\chi}\}$, and $\hat{\mathcal{J}} = \{\chi : q \in \Theta_p^{\chi}, q^\star \in \Theta_p^{\chi}\}$. The second inequality is obtained directly from the definition of the weight $c_{p,t}^{q}$. Furthermore, observing at $t = \tau + 1$, it follows that

$$\begin{aligned} c_{p,\tau+1}^{q^\star} - c_{p,\tau+1}^{q} &= \mathbb{P}_{\tau+1}(\mathbf{H}_p^{\hat{\chi}}) + \sum_{\chi \in \mathcal{J} \backslash \{\hat{\chi}\}} \mathbb{P}_{\tau+1}(\mathbf{H}_p^{\chi}) - \sum_{\chi \in \mathcal{J}'} \mathbb{P}_{\tau+1}(\mathbf{H}_p^{\chi}) \\ &= \alpha_1 \mathbb{P}_\tau(\mathbf{H}_p^{\hat{\chi}}) + \alpha_2 \sum_{\chi \in \mathcal{J} \backslash \{\hat{\chi}\}} \mathbb{P}_\tau(\mathbf{H}_p^{\chi}) - \alpha_2 \sum_{\chi \in \mathcal{J}'} \mathbb{P}_\tau(\mathbf{H}_p^{\chi}) \\ &= \alpha_2 \left(\frac{\alpha_1}{\alpha_2} \mathbb{P}_\tau(\mathbf{H}_p^{\hat{\chi}}) + \sum_{\chi \in \mathcal{J} \backslash \{\hat{\chi}\}} \mathbb{P}_\tau(\mathbf{H}_p^{\chi}) - \sum_{\chi \in \mathcal{J}'} \mathbb{P}_\tau(\mathbf{H}_p^{\chi}) \right) > 0, \end{aligned}$$

where $\alpha_1 = 1/\mathbb{P}_t(\delta_{p,t}^{\mathrm{at}} = 0)$ and $\alpha_2 = (1 - \hat{\pi}_p^{\mathrm{at}})/\mathbb{P}_t(\delta_{p,t}^{\mathrm{at}} = 0)$. The last inequality follows from the fact that $\alpha_1 > \alpha_2$ and (8.16) holds. □

Lemma 8.1 indicates that when a regular agent is connected to an adversarial agent, a detectable attack will occur almost surely. Meanwhile, Lemma 8.2 shows how a regular agent decides the connection under certain conditions. Both lemmas are used to show how the attack identification and mitigation method works.

8.6.1 The Case Where $F = 1$

When $F = 1$, there exist $|\mathcal{N}_p^{\mathrm{s}}| + 1$ hypotheses, where Θ_p^{χ}, for $\chi = 1, \ldots, |\mathcal{N}_p^{\mathrm{s}}|$, have one element. The outcome of the identification method for $F = 1$ is characterized in Proposition 8.2 as follows.

Proposition 8.2 *Suppose that Assumptions 8.1–8.3 hold, a regular agent $i \in \mathcal{R}$ applies Algorithm 8.1 with $F = 1$, and there exists an adversarial neighbor of agent p. If there is no false detection, then agent p correctly identifies the adversarial neighbor.* □

Proof A regular agent $p \in \mathcal{R}$ identifies its adversarial neighbor by concluding from the probability measures of its hypotheses. In particular, let the adversarial neighbor be denoted by $q_{\text{at}} \in \mathcal{N}_p^{\text{s}} \cap \mathcal{S}$ and the hypothesis associated to q_{at} is denoted by $\mathbf{H}_p^{\chi_{\text{at}}}$. Then, it will be shown that $\mathbb{P}_t(\mathbf{H}_p^{\chi_{\text{at}}})$ eventually becomes 1. Note that when one of the hypothesis probabilities equals 1, the others equal 0 since $\sum_{\chi=0}^{|\mathcal{N}_p^{\text{s}}|} \mathbb{P}_t(\mathbf{H}_p^{\chi}) = 1$, for any $t \in \mathbb{Z}_{\geq 0}$.

Recall that $\mathbb{P}_t(\mathbf{H}_p^{\chi})$, for all $\mathbf{H}_p^{\chi} \in \mathcal{H}_p$, evolve based on the Bayesian inference given in (8.12). From (8.13), the dynamics (8.12) can be seen as hybrid dynamics since $\delta_{p,t}^{\text{at}}, \varrho_{p,t}^{q} \in \{0, 1\}$. Note that when $\delta_{p,t}^{\text{at}} = 1$, a detectable attack occurs, whereas, when $\delta_{p,t}^{\text{at}} = 0$, no attack is detected because the adversarial neighbor either does not attack or performs an undetectable attack. Furthermore, recall also that $\varrho_{p,t}^{q} = 0$ implies agent q is blocked, so that if $q = q_{\text{at}}$, it cannot attack. Otherwise, agent q is not blocked. At each t, there is only at most one neighbor that is blocked due to constraint (8.15d) in Problem (8.15), which is solved to determine $\varrho_{p,t}$ (step 10 of Algorithm 8.1).

Now, the dynamics of all $\mathbb{P}_t(\mathbf{H}_p^{\chi})$ is analyzed based on the decision $\varrho_{p,t}$. During the period at which $\varrho_{p,t} = \mathbb{1}_{|\mathcal{N}_p^{\text{s}}|}$, the adversarial agent can attack. Note that the number of detectable attacks, $n_{p,t}^{\text{at}}$, is unbounded due to Lemma 8.1. As a result, depending on the weight γ_p, for some t where $n_{p,t}^{\text{at}}$ is sufficiently large, $\mathbb{1}_{|\mathcal{N}_p^{\text{s}}|} \notin \mathcal{Q}_{p,t}^{\star}$, e.g., see Remark 8.2. Additionally, it is observed from (8.12) and (8.13) that starting for the first time instant that $\delta_{p,t}^{\text{at}} = 1$, $\mathbb{P}_{t+1}(\mathbf{H}_p^{0}) = 0$. If agent p only has one neighbor, a detectable attack immediately leads to the identification that the neighbor is adversarial. In the following, consider the case where $|\mathcal{N}_p^{\text{s}}| > 1$ and observe the dynamics of $\mathbb{P}_t(\mathbf{H}_p^{\chi})$, for which $\mathbb{1}_{|\mathcal{N}_p^{\text{s}}|} \notin \mathcal{Q}_{p,t}^{\star}$. In particular, consider two cases: (a) when the adversarial agent is blocked and (b) when a regular neighbor is blocked.

In case (a), suppose that at time step $\underline{t}$, the adversarial agent is blocked, i.e., $\varrho_{i,\underline{t}}^{j_{\text{at}}\star} = 0$. Therefore, $\delta_{p,\underline{t}}^{\text{at}} = 0$. Moreover, according to Lemma 8.2, $\varrho_{i,\underline{t}+1}^{q_{\text{at}}\star} = 0$, implying $d_{i,\underline{t}+1}^{\text{at}} = 0$. In fact, $\varrho_{i,t}^{j_{\text{at}}\star} = 0$ and $\delta_{p,t}^{\text{at}} = 0$, for all $t \geq \underline{t}$. Hence, since $\delta_{p,t}^{\text{at}}$ and $\varrho_{p,t}$, for $t \geq \underline{t}$, are fixed, the dynamics (8.12) are smooth. By recursively applying (8.12), starting from $\underline{t}$, and considering fixed $\delta_{p,t}^{\text{at}}$ and $\varrho_{p,t}$, it follows that, for $t > \underline{t}$,

$$\mathbb{P}_t(\mathbf{H}_p^{\chi}) = \frac{(\pi_{\mathbf{H}_p^{\chi}})^{t-\underline{t}}}{\sum\limits_{\chi=0}^{|\mathcal{N}_p^{\text{s}}|} (p_{\mathbf{H}_p^{\chi}})^{t-\underline{t}} \mathbb{P}_{\underline{t}}(\mathbf{H}_p^{\chi})} \mathbb{P}_{\underline{t}}(\mathbf{H}_p^{\chi}), \ \forall \mathbf{H}_p^{\chi} \in \mathcal{H}_p,$$

where $\pi_{\mathbf{H}_p^\chi} = \mathbb{P}_t(\delta_{p,t}^{\text{at}} = 0|\mathbf{H}_p^\chi)$, i.e.,

$$\pi_{\mathbf{H}_p^\chi} = \begin{cases} 1, & \text{for } \chi \in \{0, \chi_{\text{at}}\}, \\ 1 - \hat{\pi}_p^{\text{at}}, & \text{otherwise.} \end{cases}$$

Hence, for $\mathbf{H}_p^{\chi_{\text{at}}}$,

$$\mathbb{P}_t(\mathbf{H}_p^{\chi_{\text{at}}}) = \frac{1}{\mathbb{P}_{\underline{t}}(\mathbf{H}_p^{\chi_{\text{at}}}) + \sum\limits_{\chi \neq m_{\text{at}}} (p_{\mathbf{H}_p^\chi})^{t-\underline{t}} \mathbb{P}_{\underline{t}}(\mathbf{H}_p^\chi)} \mathbb{P}_{\underline{t}}(\mathbf{H}_p^{\chi_{\text{at}}}).$$

Thus, it holds that $\lim_{t\to\infty} \mathbb{P}_t(\mathbf{H}_p^{\chi_{\text{at}}}) = 1$, since $\pi_{\mathbf{H}_p^\chi} < 1$ for all $\chi \neq \chi_{\text{at}}$. Furthermore, it also holds that $\lim_{t\to\infty} \mathbb{P}_t(\mathbf{H}_p^\chi) = 0$ for $\chi \neq \chi_{\text{at}}$.

In case (b), suppose that at time step $\underline{t}$, a regular neighbor $q \in \mathcal{N}_p^{\text{s}} \backslash \{q_{\text{at}}\}$ is blocked, i.e., $\varrho_{p,\underline{t}}^{q\star} = 0$. According to Lemma 8.2 and the dynamics (8.12) and (8.13), the neighbor q is blocked as long as $\delta_{p,t}^{\text{at}} = 0$, for $t \geq \underline{t}$. However, consider that at some $\bar{t} \geq \underline{t}$, $\delta_{p,\bar{t}}^{\text{at}} = 1$. Then, based on (8.12) and (8.13), $\mathbb{P}_{\bar{t}+1}(\mathbf{H}_p^{\chi_q}) = 0$. Note that since q_{at} is not blocked for $\underline{t} \leq k \leq \bar{t}$, a detectable attack will occur almost surely (Lemma 8.1). Thus, if any regular neighbor $q \in \mathcal{N}_p^{\text{s}} \backslash \{q_{\text{at}}\}$ is being blocked, its probability $\mathbb{P}_t(\mathbf{H}_p^{\chi_q})$ will eventually become 0. As a result, if another regular neighbor is blocked at $t = \bar{t} + 1$, case (b) repeats, whereas if the adversarial neighbor is blocked, case (a) follows. □

Remark 8.5 Based on Proposition 8.2, each regular agent that applies Algorithm 8.1 can identify its adversarial neighbor correctly for the case $F = 1$ provided that there is no false detection. Based on Definition 8.6, a false detection occurs when the uncertainty from the load, $w_{i,t}^{\text{d}}$, is outside of the probabilistic bounds. Since the computed decision is an ε-level solution (Proposition 8.1), the probability of $w_{i,t}^{\text{d}}$ being outside of the bounds is at most ε, which can be set to be small. Nevertheless, if false detections are made during the operation, then all hypothesis probabilities will eventually become zero. When they are all zero, the identification process can then be restarted. □

8.6.2 The Case Where $F > 1$

In the case that $F > 1$, a regular agent might not be able to identify the adversarial neighbors. However, it can block all adversarial neighbors, as presented in Proposition 8.3.

Proposition 8.3 *Suppose that Assumptions 8.1–8.3 hold, regular agent $p \in \mathcal{R}$ applies Algorithm 8.1 with $F > 1$, and there exist at most F adversarial neighbors of agent $p \in \mathcal{R}$. If there is no false detection, then the hypothesis probability associated* *to one of the sets of neighbors that contain all adversarial neighbors converges to one and, when all hypothesis probabilities* $\mathbb{P}_t(\mathbf{H}_p^{\chi})$ *have converged, agent p blocks all adversarial neighbors.* □

Proof The lines of proof are similar to those of Proposition 8.2. Let $\mathcal{S}_p$ denote the set of adversarial neighbors of agent p, i.e., $\mathcal{S}_p = \mathcal{N}_p^{\mathrm{s}} \cap \mathcal{S}$. Note that by Definition 8.2, $|\mathcal{S}_p| \le F$. The analysis is based on the dynamics of $\mathbb{P}_t(\mathbf{H}_p^{\chi})$, for all $\mathbf{H}_p^{\chi} \in \mathcal{H}_p$, at $t \ge t_0$, for which $\mathbb{1}_{|\mathcal{N}_p^{\mathrm{s}}|} \notin \mathcal{Q}_{p,t}^{\star}$. Firstly, observe that, if $F \ge |\mathcal{N}_p^{\mathrm{s}}|$, then there are only two hypotheses. The assertion immediately holds since when there is a detectable attack, $\mathbb{P}(\mathbf{H}_p^0) = 0$. Furthermore, since $\mathbb{1}_{|\mathcal{N}_p^{\mathrm{s}}|} \notin \mathcal{Q}_{p,t}^{\star}$, all neighbors are disconnected. Now observe two possible cases on the decision of $\varrho_{p,t}$ for $F < |\mathcal{N}_p^{\mathrm{s}}|$. The cases are similar to those that are explained in the proof of Proposition 8.2.

In case (a), suppose that at $\underline{t}$, agent p blocks all adversarial neighbors, $q \in \mathcal{S}_p$, i.e., $\varrho_{p,\underline{t}}^{q} = 0$ for all $q \in \mathcal{S}_p$. Since agent p is disconnected from all adversarial neighbors, at $t = \underline{t}$, $\delta_{p,\underline{t}}^{\mathrm{at}} = 0$. Furthermore, based on Lemma 8.2, the decision to block all adversarial neighbors hold for $t \ge \underline{t}$. Therefore, $\mathbb{P}_t(\mathbf{H}_p^{\chi})$, for all $\chi = 0, 1, \ldots, n_p^{\mathrm{H}}$, evolve smoothly for $t \ge \underline{t}$. Let $\Theta_{p,\underline{t}}^{\hat{\chi}}$ denotes the set of neighbors that are disconnected by agent p at time $\underline{t}$, i.e., $\Theta_{p,\underline{t}}^{\hat{\chi}} = \{q \in \mathcal{N}_p^{\mathrm{s}} : \varrho_{p,\underline{t}}^{j} = 0\}$ and $\mathbf{H}_p^{\hat{\chi}}$ denotes the hypothesis associated to $\Theta_{p,\underline{t}}^{\hat{\chi}}$. Note that $\mathcal{S}_p \subseteq \Theta_{p,\underline{t}}^{\hat{\chi}}$, where the equality holds if $|\mathcal{S}_p| = F$. By evaluating the dynamics (8.12) similarly to the proof of Proposition 8.2, it holds that $\lim_{t\to\infty} \mathbb{P}_t(\mathbf{H}_p^{\hat{\chi}}) = 1$ whereas the other hypothesis probabilities converge to 0.

In case (b), suppose that at $\underline{t}$, some adversarial neighbors are not blocked, i.e., the set $\{q \in \mathcal{S}_p : j \notin \Theta_{p,\underline{t}}^{\hat{\chi}}\}$ is nonempty. As in the case $F = 1$, if there is no attack, due to Lemma 8.2, the set of neighbors $\Theta_{p,\underline{t}}^{\hat{\chi}}$ will still be blocked at the next time step, $t = \underline{t} + 1$. Furthermore, since $\mathbb{P}_t(\mathbf{H}_p^{\chi})$ for all $\chi = 1, \ldots, n_p^{\mathrm{H}}$ are initialized equally, only $\mathbb{P}_t(\mathbf{H}_p^{\hat{\chi}})$ evolves differently, for $t \ge \underline{t}$. The other probabilities, $\mathbb{P}_t(\mathbf{H}_p^{\chi})$, for $\chi \neq \hat{\chi}$ such that $\mathbb{P}_t(\mathbf{H}_p^{\chi}) \neq 0$, are equal, for $t \ge \underline{t}$, since they are multiplied by the same factor, which is either $(1 - \hat{\pi}_p^{\mathrm{at}})/\mathbb{P}(\delta_{p,t}^{\mathrm{at}})$ when there is no attack, or $\hat{\pi}_p^{\mathrm{at}}/\mathbb{P}(\delta_{p,t}^{\mathrm{at}})$ when there is an attack. Nevertheless, the occurrence of the next attack is with probability 1 (Lemma 8.1). Suppose that the next attack occurs at $\bar{t}$. Therefore, $\mathbb{P}_{\bar{t}}(\mathbf{H}_p^{\hat{\chi}}) = 0$, while other hypotheses that have probability strictly larger than zero at $t = \bar{t} - 1$, have an equal value at $t = \bar{t}$, denoted by $\pi_{\bar{t}}$. Note that the number of these hypotheses is $1/\pi_{\bar{t}}$. The decision $\varrho_{p,\bar{t}+1}$ depends on solving Problem (8.15) and step 4 in Algorithm 8.1. Due to step 4 in Algorithm 8.1, a different set of neighbors, i.e., $\Theta_{p,\bar{t}+1}^{\hat{\chi}} \neq \Theta_{p,\underline{t}}^{\hat{\chi}}$, is disconnected at $t = \bar{t} + 1$. If $\mathcal{S}_p \subseteq \Theta_{p,\bar{t}+1}^{\hat{\chi}}$, case (a) will follow, otherwise case (b) will be repeated. Note that, since the number of sets of disconnected neighbors that do not include all adversarial agents is limited and such a set cannot be chosen twice, eventually a set of neighbors, Θ_p^{χ}, which includes all adversarial neighbors, i.e., $\mathcal{S}_p \subseteq \Theta_p^{\chi}$, is disconnected. Thus, eventually case (a) occurs. □

Remark 8.6 Proposition 8.3 shows that each regular agent that applies Algorithm 8.1 can eventually block all adversarial neighbors for any $F \ge 1$, provided that there is no false detection. Therefore, similar to the case $F = 1$, setting a small desired level of violation ε implies a high probability of blocking all adversarial neighbors. Furthermore, when false detections occur, then the identification and mitigation process can be restarted after all hypothesis probabilities have become zero. □

8.7 Case Study

In order to show the effectiveness of the proposed approach in mitigating attacks and identifying adversarial agents, a simplified model of the benchmark case PG&E 69-bus distribution network is considered. Using the partition described in Sect. 2.4, each agent is considered to have one (supra-)node, which contains dispatchable generation and storage units as well as loads. The network of the agents, $\mathcal{G}_t^s$, which is fixed for all $t \in \mathbb{Z}_{\geq 0}$, is shown in Fig. 8.1. Furthermore, the parameter values of the components of each subsystem are given in Table 8.1. It is considered that there are two types of load profiles, which are residential and industrial loads. Moreover, suppose that subsystems 1, 2, 5, and 6 have industrial load profiles whereas the others have residential profiles. In addition, it is set that agents 2, 6, and 7 are adversarial.

Two simulation studies are carried out. The first study shows how the overall scheme works in two scenarios, which correspond to the conditions in Propositions 8.2 and 8.3, while the second study emphasizes on the attack detection scheme. The simulations are carried out in MATLAB with YALMIP [20] using a computer with 2.6 GHz Intel Core i7 CPU and 16 GB of RAM. In addition, it is set that the sampling time is 15 min, the prediction horizon is four time steps.

8.7.1 *Performance of the Overall Scheme*

In this simulation, where the simulation time is one day (96 steps), the adversarial agents, $p = 2, 6, 7$, attack with $\pi_p^{\text{at}} = 0.3$. The industrial and the residential load profiles are based on realistic data collected in a large-scale study [21, 22]. Furthermore, the network has solar-based energy sources, the profiles of which are also based on realistic data [22]. The attack strategy of the adversarial agents is to reduce the production of their dispatchable generation units randomly. Here, Algorithm 8.1 is applied to the previously described system. Two simulation scenarios are considered, where it is assumed that false detection never occurs. In the first scenario, it is assumed that $F = 1$, whereas in the second scenario $F = 2$. Figures 8.2 and 8.4 show

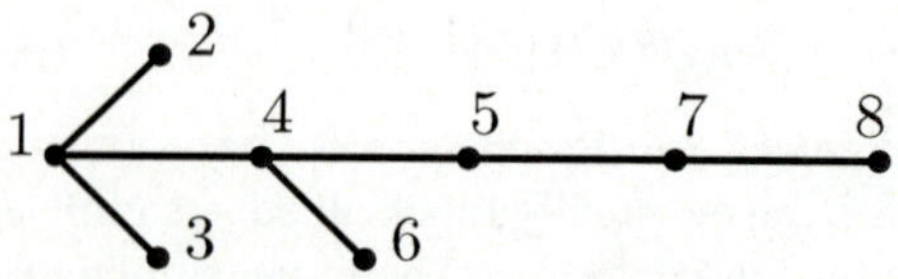

Fig. 8.1 The network of supra-nodes/agents obtained from the benchmark case

Table 8.1 Parameters of the agents

Parameters	Value	Unit	Node (i)
$x_i^{\min}, x_i^{\max}, x_{i,0}$	30, 80, 50	%	All
$u_i^{\text{ch}}, u_i^{\text{dh}}$	300, 300	kW	All
$u_i^{\text{dg,min}}, u_i^{\text{dg,max}}$	0, 1000	kW	3, 4, 7, 8
	0, 2000		1, 2, 5, 6
$v_i^{j,\max}$, for all $j \in \mathcal{N}_i$	110	kW	All
$u_i^{\text{tp,max}}$	2000	kW	1
	0		Else
$e_{\text{cap},i}$	500	kWh	3, 4, 7, 8
	1000		1, 2, 5, 6
a_i	0.98	–	All
$c_i^{\text{st}}, c_i^{\text{tp}}, c_i^{\text{t}}$	1, 250, 0.1	–	All
c_i^{dg}	5	–	2, 4, 6, 7
	10		1, 3, 5, 8
ε_i	0.01	–	All

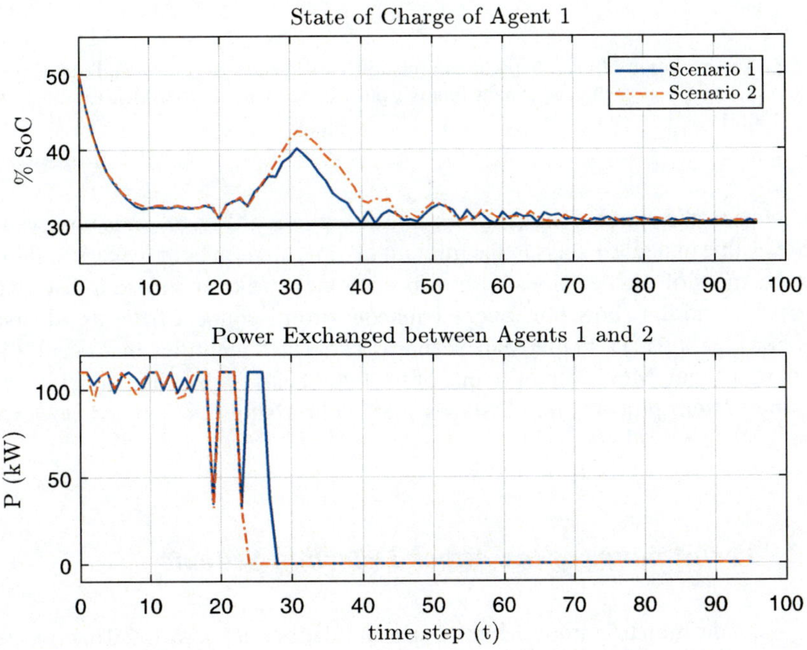

Fig. 8.2 The evolution of SoC of agent 1 (top plot) and the power exchanged between agent 1 and its adversarial neighbor, agent 2 (bottom plot)

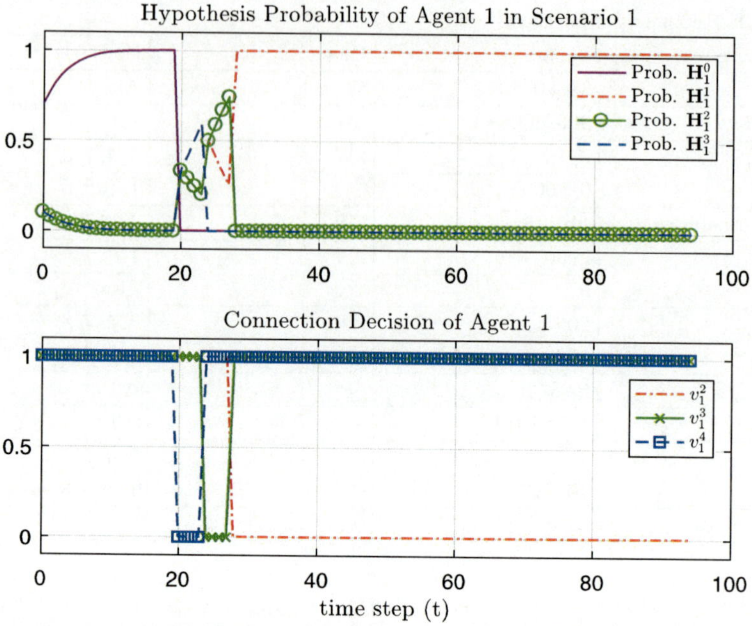

Fig. 8.3 The evolution of each hypothesis probability of agent 1 (top plot) and the connection decision of agent 1 at each time instant (bottom plot) in Scenario 1. Note that $\Theta_1^1 = \{2\}$, $\Theta_1^2 = \{3\}$, $\Theta_1^3 = \{4\}$

some plots of the simulation results. From the top plot of Fig. 8.2, it is observed that the SoC value of agent 1 stays in the limit for all time steps in both scenarios, showing the robustness of the decisions with respect to the attacks and system disturbance. Figures 8.3 and 8.4 show how agent 1 manages to disconnect from its adversarial neighbor (agent 2). Particularly in Scenario 1, agent 1 identifies that agent 2 is an adversarial agent. Moreover, once one of the hypothesis probability values converge to 1, the bottom plot of Fig. 8.2 shows that agent 1 stops exchanging power with agent 2.

8.7.2 *Performance of the Attack Detection Scheme*

In the second simulation study, Monte Carlo simulations are performed to observe the attack detection scheme. Suppose that the adversarial agents $p = 2, 6, 7$ attack with $\pi_p^{\text{at}} = 0.5$ and the regular agents do not apply the active strategy of disconnecting their neighbors so that the regular agents are always subject to attacks. Note that the probability of attack is set to be quite high in order to observe more attacks, particularly undetectable ones. For these simulations, load and renewable power generation profiles from the first simulation study are perturbed. Seven simulations,

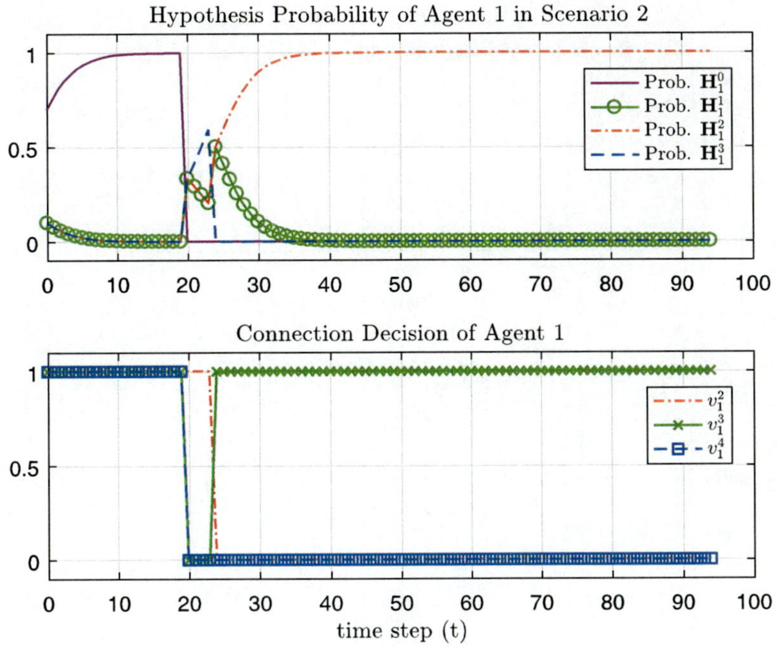

Fig. 8.4 The evolution of each hypothesis probability of agent 1 (top plot) and the connection decision of agent 1 at each time instant (bottom plot) in Scenario 2. Note that $\Theta_1^1 = \{2, 3\}$, $\Theta_1^2 = \{2, 4\}$, $\Theta_1^3 = \{3, 4\}$

Table 8.2 Detection results of the regular agents

Agent	Correct detection (%)	Undetectable attack (%)	False detection (%)
1	82.74	16.82	0.44
3	99.40	0.00	0.60
4	82.59	17.11	0.30
5	79.61	20.09	0.30
8	49.85	50.15	0.00
Total	78.84	20.83	0.33

each of which is 96 time steps length, are carried out. Therefore, 672 detection instants, for each regular agent, can be observed. Table 8.2 shows the summary of the attack detection outcomes whereas Fig. 8.5 shows the attack detection of agent 1 in some time steps of one simulation. Correct detection means that an agent correctly detects whether there is an attack or not. As expected from the stochastic method, the probability of false detection is less than ε. One can see a false detection in the top plot of Fig. 8.5 at time step $t = 7$.

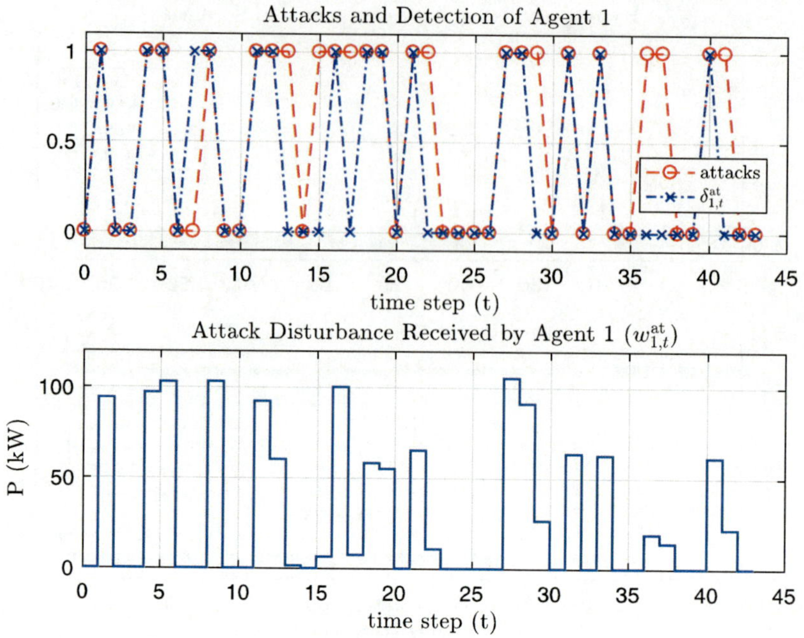

Fig. 8.5 Top plot shows the detection variable $\delta^{at}_{1,t}$ (dot-dashed blue plot with cross markers) and the actual event, i.e., whether there is an attack (1) or not (0) (dashed red plot with circle markers). Bottom plot shows the attack disturbance of agent 1, $w^{at}_{1,t}$. These plots are taken from one simulation

8.7.3 Discussions

The local performance improvements obtained by each adversarial agent when performing successful attacks in both simulation studies presented in Sects. 8.7.1 and 8.7.2 are shown in Table 8.3. Furthermore, as can be seen in Table 8.3, the performance of the regular agents are degraded by the attacks. The degradation is relatively low compared to the improvement obtained by the adversarial agents since most of the attacks can still be handled by the storage unit, whose cost per unit is cheaper than using the dispatchable generator, whereas the adversarial agents manage to reduce the power production of their generators by performing the attacks. However, note that the available power in the storage unit of a regular agent is lower after an

Table 8.3 Average local performance change during the attacks

Case study	Adversarial agent (%)			Regular agent (%)			
	2	6	7	1	4	5	8
8.7.1	19.7	1.5	4.5	0.0	−0.3	−0.1	−0.2
8.7.2	35.2	8.9	53.3	−0.4	−1.1	−0.6	−3.7

Note Agent 3 does not have any adversarial neighbors

attack occurs. Therefore, the performance of a regular agent after an attack might still deteriorate and this performance degradation is not captured in Table 8.3.

It is also worth mentioning that an adversarial agent might actually perform an undetectable attack. Although the performance improvement obtained by an undetectable attack is less than a detectable one since undetectable attack is limited, the identification process will take longer time to finish. For instance, as can be seen in Table 8.2, the probability of correct detections of agent 8 is 49.85%, implying that the adversarial neighbor of agent 8 has successfully performed undetectable attacks half of the time. Note that, in the case study of Sect. 8.7.2, the average performance improvements obtained by the adversarial agents 2, 6, and 7 by performing undetectable attacks are lower than the total average shown in Table 8.3 (9.7%, 0.1%, and 52.61%, respectively). Similarly, the performance degradation of each regular agent by undetectable attacks is also quite low (less than 0.6%). Nevertheless, in order to guarantee that the performed attack is undetectable, an adversarial agent will require local information of its neighbors, which is not shared in the presented problem setting. Therefore, it is difficult for the adversarial agents to keep attacking without being detected.

8.8 Summary

A distributed approach for the economic dispatch problem of energy systems in the presence of adversaries has been proposed. The adversarial actions are considered as uncertain disturbances and the economic dispatch problem is formulated as a chance-constrained problem. Thus, a two-step stochastic approach is applied so that the control inputs computed are robustly feasible against the adversarial behavior. Furthermore, the proposed approach also includes a methodology to identify the adversarial agents and mitigate the attacks from these agents. The methodology is based on hypothesis testing using Bayesian inference and requires each regular agent to solve a local mixed-integer problem to decide the connection with its neighbors. Therefore, the proposed scheme is a combination of active and passive methods to deal with unexpected disturbances.

References

1. Velarde P, Valverde L, Maestre JM, Ocampo-Martinez C, Bordons C (2017) On the comparison of stochastic model predictive control strategies applied to a hydrogen-based microgrid. J Power Sources 343:161–173
2. Sharma DD, Singh SN, Lin J, Foruzan E (2017) Agent-based distributed control schemes for distributed energy storage systems under cyber attacks. IEEE Trans Emerg Sel Topics Circuits Syst 7(2):307–318
3. Pasqualetti F, Dörfler F, Bullo F (2013) Attack detection and identification in cyber-physical systems. IEEE Trans Autom Control 58(11):2715–2729

4. LeBlanc HJ, Zhang H, Koutsoukos X, Sundaram S (2013) Resilient asymptotic consensus in robust networks. IEEE J Sel Areas Commun 31(4):766–781
5. Dibaji SM, Ishii H, Tempo R (2018) Resilient randomized quantized consensus. IEEE Trans Autom Control 63(8):2508–2522
6. Feng Z, Wen G, Hu G (2017) Distributed secure coordinated control for multiagent systems under strategic attacks. IEEE Trans Cybern 47(5):1273–1284
7. Velarde P, Maestre JM, Ishii H, Negenborn RR (2017) Scenario-based defense mechanism for distributed model predictive control. In: Proceedings of the IEEE 56th conference on decision and control (CDC), Melbourne, Australia, 2017, pp 6171–6176
8. Tanaka T, Gupta V (2016) Incentivizing truth-telling in MPC-based load frequency control. In: Proceedings of the IEEE 55th conference on decision and control (CDC), Las Vegas, USA, 2016, pp 1549–1555
9. Ananduta W, Maestre JM, Ocampo-Martinez C, Ishii H (2018) Resilient distributed energy management for systems of interconnected microgrids. In: Proceedings of the IEEE 57th conference on decision and control (CDC), Miami, USA, 2018, pp 6159–6164
10. Ananduta W, Maestre JM, Ocampo-Martinez C, Ishii H (2019) A resilient approach for distributed MPC-based economic dispatch in interconnected microgrids. In: Proceedings of the European control conference, Naples, Italy, 2019, pp 891–896
11. Ananduta W, Maestre JM, Ocampo-Martinez C, Ishii H (2020) Resilient distributed model predictive control for energy management of interconnected microgrids. Optim Control Appl Methods 41(1):146–169
12. Schwaegerl C, Tao L (2013) The microgrids concept, chapter 1. Wiley, pp 1–24
13. Margellos K, Rostampour V, Vrakopoulou M, Prandini M, Andersson G, Lygeros J (2013) Stochastic unit commitment and reserve scheduling: a tractable formulation with probabilistic certificates. In: Proceedings of European control conference, 2013, pp 2513–2518
14. Vrakopoulou M, Margellos K, Lygeros J, Andersson G (2013) A probabilistic framework for reserve scheduling and $N-1$ security assessment of systems with high wind power penetration. IEEE Trans Power Syst 28(4):3885–3896
15. Margellos K, Goulart P, Lygeros J (2014) On the road between robust optimization and the scenario approach for chance constrained optimization problems. IEEE Trans Autom Control 59(8):2258–2263
16. Calafiore GC, Campi MC (2006) The scenario approach to robust control design. IEEE Trans Autom Control 51(5):742–753
17. Guerrero JM, Vasquez JC, Matas J, de Vicuna LG, Castilla M (2011) Hierarchical control of droop-controlled ac and dc microgrids—a general approach toward standardization. IEEE Trans Ind Electron 58(1):158–172
18. Ingimundarson A, Bravo JM, Puig V, Alamo T, Guerra P (2008) Robust fault detection using zonotope-based set-membership consistency test. Int J Adapt Control Signal Process 23(4):311–330
19. Capinski M, Kopp PE (2013) Measure, integral and probability. Springer
20. Löfberg J (2004) YALMIP: a toolbox for modeling and optimization in MATLAB. In: Proceedings of the CACSD conference, Taipei, Taiwan, 2004, pp 284–289
21. Hayashi Y, Fujimoto Y, Ishii H, Takenobu Y, Kikusato H, Yoshizawa S, Amano Y, Tanabe SI, Yamaguchi Y, Shimoda Y, Yoshinaga J, Watanabe M, Sasaki S, Koike T, Jacobsen HA, Tomsovic K (2018) Versatile modeling platform for cooperative energy management systems in smart cities. Proc IEEE 106(4):594–612
22. New Energy Industrial Technology Development Organization (NEDO) Demonstrative research on grid-interconnection of clustered photovoltaic power generation systems

Chapter 9
Concluding Remarks

9.1 Contributions

The main objectives of this thesis are to design non-centralized MPC-based schemes for economic dispatch of large-scale energy system and to study the communication and cooperation processes of the schemes. Motivated by the fact that energy systems are critical infrastructures, potential communication and cooperation problems that might occur during the implementation of the schemes are investigated and methodologies to cope with them are developed. The contributions of this thesis are summarized as follows.

The first contribution is the development of two distributed optimization methods, which work under an imperfect communication process, to solve edge-based resource sharing problems, which include the economic dispatch problem of energy systems formulated in Chap. 2. The methods, which are called the DAL and ADMM-based methods, are based on the augmented Lagrangian approach and introduced in Chap. 3. Then, in Chap. 5, the imperfect communication process is modeled as a random time-varying network and random asynchronous updates. Considering this model, the DAL and ADMM-based methods are modified and their convergence properties are investigated. Finally, Chap. 6 showcases the implementation of the methods under the MPC-based framework for the energy management of large-scale systems through simulations of a benchmark case.

Considering the importance of the communication infrastructure in DMPC schemes, a consensus-based information-exchange protocol is proposed as the second contribution of the thesis, which is provided in Chap. 4. The aim of this proposal is to improve the resiliency of the communication infrastructure with respect to failures. As opposed to the first contribution, which focuses on a specific distributed approach, the second contribution can be applied to a broad class of DMPC strategies.

Furthermore, in effort to obtain a scheme that has less reliance on the exchanges of information, a coalition-based economic dispatch scheme that uses an online system repartitioning is developed. By properly partitioning the network, self-sufficient coalitions of subsystems with local economic dispatch subproblems, can be obtained. To this end, an event-triggered repartitioning procedure, a coalition-formation procedure, and a coalition-based economic dispatch scheme are proposed as the third contribution, which is presented in Chap. 7.

W. W. Ananduta, *Non-centralized Optimization-Based Control Schemes for Large-Scale Energy Systems*, Springer Theses,
https://doi.org/10.1007/978-3-030-89803-8_9

Finally, one potential cooperation problem in the scheme, namely the non-compliance of agents, is studied in Chap. 8. As the last contribution, a methodology to cope with non-compliance problems during the decision implementation phase is proposed and analyzed. The methodology includes the robustification of the decisions and the identification and mitigation of the attack.

9.2 Answering Research Questions

To conclude the discussion presented in Chaps. 2–8, the key research questions formulated in Sect. 1.3 are addressed as follows:

(Q_1) *How to design and implement a non-centralized MPC-based economic dispatch scheme for large-scale energy systems?*
An MPC-based economic dispatch for large-scale energy systems, as shown in Algorithm 2.1, mainly consists of two tasks: solving a finite-time horizon optimization problem at each time step and implementing the first decisions, i.e., those that correspond to the current time step. Clearly, prior to designing methodologies to perform these tasks, a mathematical problem formulation is required. Section 2.2 formulates the economic dispatch problem as a convex problem with edge-based coupling constraints.

In a non-centralized scheme, the first task, i.e., solving the formulated optimization problem, must be carried out by local controllers. In this regard, two distributed algorithms, namely the DAL and ADMM-based methods, which are suitable to solve the problem have been proposed in Chap. 3. The algorithms are based on the augmented Lagrangian approach and exploit the structure of the coupling constraints. These algorithms are iterative and require information exchanges among local controllers at each iteration. As discussed in Chap. 6, some techniques, such as the Nesterov's acceleration and the warm start methods, can be applied to reduce the number of iterations. Furthermore, it has also been shown that the sequence produced by any of these algorithms converges to an optimal solution to the economic dispatch problem.

Moreover, with the objective of reducing communication flow, an alternative non-centralized scheme has also been proposed in Chap. 7. The scheme is based on an online system partitioning and coalition formation. When the resulting partitions are self-sufficient, the economic dispatch problem can even be solved in a decentralized manner without requiring communication and iterations.

It is worth mentioning that the cooperation and communication of the local controllers play an important role in a non-centralized control scheme. Furthermore, there might be potential problems with the communication and cooperation that must be handled and are addressed by the remaining key research questions 1.3–1.3.

(Q_2) *How to design an information-exchange protocol for distributed MPC approaches that is resilient against communication failures?*
The resiliency of the information exchange process, which is an integral part of a distributed MPC strategy, can be improved by applying a consensus algorithm. By using such a procedure, the required information does not always have to come directly from its source. Therefore, the distributed MPC algorithm can still be performed even though some communication links fail. As presented in Chap. 4, a consensus-based information-exchange protocol is designed. Moreover, a method to partition the communication network and that to reconfigure the partitions are also presented.

(Q_3) *Do distributed optimization algorithms work over time-varying communication networks and in an asynchronous manner?*
It has been shown in Chap. 5 that the DAL and ADMM-based methods produce a sequence that almost surely converges to an optimal solution of the economic dispatch problem when the communication network is stochastically time-varying and the updates are performed asynchronously. In order to do so, the distributed algorithms must be adjusted. Nevertheless, the adjustment is intuitive in the sense that some variables are updated based on the availability of new information. It is worth discussing that some references mentioned in Chap. 5 discuss different distributed optimization methods that work over random communication networks or asynchronous updates. There are also some other distributed approaches that work over time-varying network under the jointly connected assumption. Although Chap. 5 does not give a general answer to the question, it provides some examples of distributed methods that work under the required conditions and a suggestion of how other distributed algorithms might be modified such that they work under imperfect communication conditions.

(Q_4) *How to perform online partitioning on large-scale energy systems in a distributed manner?*
Firstly, the task of decomposing the system is translated into a graph partitioning problem. Although vast options of graph partitioning methods are available in the literature, one cannot randomly choose any of these methods. Typically, the way the network is partitioned must be in accordance with the partitioning objective. Moreover, the requirement that it must be performed in a distributed manner also gives an additional challenge. Chapter 7 provides a distributed repartitioning procedure, which can be classified as a local improvement method and is particularly designed with the objective of solving the economic dispatch problem with a non-centralized scheme. Additionally, as shown in Chap. 7, the partitioning solution of the method converges to a local optimum.

(Q_5) *How can online partitioning be used in a non-centralized economic dispatch scheme of large-scale energy systems?*
A large-scale energy system can be partitioned into self-sufficient subsystems,

which are able to meet their demand locally without relying on importing energy from other subsystems or a third party. By having a group of self-sufficient subsystems, the economic dispatch problem of the network can then be decomposed into local subproblems of the subsystems. However, due to the time-varying nature of the loads and non-dispatchable power generation, self-sufficiency of the subsystems might be compromised. Therefore, the partitions of the network must be adjusted and an online partitioning scheme is required to make the modification, as shown in Chap. 7. Additionally, the partitioning scheme proposed in Chap. 7 does not guarantee that the partitions are always self-sufficient. Therefore, the subsystems that are not self-sufficient must form a coalition with some of their neighbors to ensure that the local economic dispatch subproblem that they solve is feasible. To that end, a coalition formation method is also proposed.

(Q_6) *How to design a non-centralized MPC-based economic dispatch scheme that is resilient against non-compliance issues?*
The non-compliance problem, where some adversarial agents do not cooperate and selfishly operate at their own benefits by not implementing the computed solution, is dealt with by two mechanisms. Firstly, the decisions of each regular agent are robustified such that the adversarial behavior does not affect the operation of the regular agents, e.g., violating the operational constraints. This passive mechanism uses a stochastic method where the adversarial behavior is considered as a source of uncertainty. Then, a probabilistic bound of the uncertainties that influence the regular agents is computed and a robust economic dispatch problem is formulated. Secondly, the adversarial agents are at least localized. To that end, a hypothesis testing method based on Bayesian inference is employed to identify the adversarial agents. In a specific case, it is shown that the adversarial agents are identified. These two mechanisms are effective for systems of interconnected microgrids, which can operate on the island and connected modes, as presented in Chap. 8.

9.3 Directions of Future Research

This thesis has investigated some communication and cooperation problems on the non-centralized MPC schemes, mainly for the economic dispatch problem of large-scale energy systems. However, each of the methodologies that have been developed still requires further improvements and investigation. Hence, some ideas for future research are outlined as follows:

- The stochastic distributed optimization methods presented in Chap. 5 might be extended to problems with global (coupled) cost functions or non-convex coupling constraints, which are relevant to the energy management and the optimal power flow problems of electrical energy systems. Furthermore, the convergence analysis of the accelerated versions of the methods, which are shown in Chap. 6, needs to be done.

- As noted in Remark 3.2, the solutions to the local optimization problems in the proposed distributed optimization methods can be computed by efficient solvers. However, the methods might be extended by considering an inexact optimization approach, such as first-order methods discussed in [1], in order to reduce the computational complexity or to deal with non-dual-friendly functions. In this regard, further analysis of the performance of this possible extension of the method is required.
- Related to the information-exchange protocol in Chap. 4, an automatic and systematic technique to design an optimal communication network, which is also resilient, is required to complement the protocol. Furthermore, a time-varying partitioning method may improve the proposed reconfiguration procedure.
- Since the coalition-based economic dispatch scheme provides a trade-off between suboptimality and communication burden, it might be better to include the suboptimality measure as another criterion that not only triggers but also becomes the objective of the repartitioning. In this regard, the non-centralized control scheme proposed in Chap. 7, including the repartitioning and coalition formation methods, need to be modified.
- On the topic of cooperation problem of non-centralized MPC schemes, it is important to analyze potential clever attacks that are hard to be detected. In this regard, the interaction between regular and adversarial agents might be analyzed using game theoretic approaches. Furthermore, there is still a room for improvement of the identification method presented in Sect. 8.4. For instance, by employing an information-sharing scheme of the hypothesis probability among the agents.
- There are other communication and cooperation problems that might occur and must be dealt with, such as noisy communication and adversarial behaviors when solving the optimization problem. Addressing these problems might require different tools and techniques than those that are used in this thesis.

Reference

1. Uribe CA, Lee S, Gasnikov A, Nedić A (2018) A dual approach for optimal algorithms in distributed optimization over networks. arXiv:1809.00710, pp 1–37

Appendix A
Proofs

A.1 Proof of Lemma 3.2

Recall the result in Lemma 3.1. In particular, rearrange (3.12) and add the term $2\sum_{i\in\mathcal{N}}\sum_{j\in\mathcal{N}_i}\langle\hat{\boldsymbol{v}}_i^j(k)-\boldsymbol{v}_i^j(k), \hat{\boldsymbol{v}}_i^j(k)-\boldsymbol{v}_i^{j\star}\rangle$ on both side of the inequality. It is obtain that

$$
\begin{aligned}
&\sum_{i\in\mathcal{N}}\sum_{j\in\mathcal{N}_i}\Big(\langle\boldsymbol{\lambda}_i^j(k)-\boldsymbol{\lambda}_i^{j\star}, \hat{\boldsymbol{v}}_i^j(k)+\hat{\boldsymbol{v}}_j^i(k)\rangle+2\langle\hat{\boldsymbol{v}}_i^j(k)-\boldsymbol{v}_i^j(k), \hat{\boldsymbol{v}}_i^j(k)-\boldsymbol{v}_i^{j\star}\rangle\Big)\\
&\leq\sum_{i\in\mathcal{N}}\Big(-\sigma_i\|\boldsymbol{u}_i(k+1)-\boldsymbol{u}_i^\star\|_2^2-\sum_{j\in\mathcal{N}_i}\|\hat{\boldsymbol{v}}_i^j(k)+\hat{\boldsymbol{v}}_j^i(k)\|_2^2\\
&\qquad+2\sum_{j\in\mathcal{N}_i}\Big(\langle\hat{\boldsymbol{v}}_j^i(k)-\boldsymbol{v}_j^i(k), \hat{\boldsymbol{v}}_i^j(k)-\boldsymbol{v}_i^{j\star}\rangle\langle\hat{\boldsymbol{v}}_i^j(k)-\boldsymbol{v}_i^j(k), \hat{\boldsymbol{v}}_i^j(k)-\boldsymbol{v}_i^{j\star}\rangle\Big)\Big).
\end{aligned}
\tag{A.1}
$$

The second inner product in the summation on the left-hand side of the inequality can be expressed as follows:

$$
\begin{aligned}
\langle\hat{\boldsymbol{v}}_i^j(k)-\boldsymbol{v}_i^j(k), \hat{\boldsymbol{v}}_i^j(k)-\boldsymbol{v}_i^{j\star}\rangle &= \langle\hat{\boldsymbol{v}}_i^j(k)-\boldsymbol{v}_i^j(k), \hat{\boldsymbol{v}}_i^j(k)-\boldsymbol{v}_i^{j\star}+\boldsymbol{v}_i^j(k)-\boldsymbol{v}_i^j(k)\rangle\\
&= \langle\hat{\boldsymbol{v}}_i^j(k)-\boldsymbol{v}_i^j(k), \boldsymbol{v}_i^j(k)-\boldsymbol{v}_i^{j\star}\rangle+\|\hat{\boldsymbol{v}}_i^j(k)-\boldsymbol{v}_i^j(k)\|_2^2.
\end{aligned}
$$

Moreover, the last two terms on the right-hand side of (A.1), i.e.,

$$
2\sum_{i\in\mathcal{N}}\sum_{j\in\mathcal{N}_i}\Big(\langle\hat{\boldsymbol{v}}_j^i(k)-\boldsymbol{v}_j^i(k), \hat{\boldsymbol{v}}_i^j(k)-\boldsymbol{v}_i^{j\star}\rangle+\langle\hat{\boldsymbol{v}}_i^j(k)-\boldsymbol{v}_i^j(k), \hat{\boldsymbol{v}}_i^j(k)-\boldsymbol{v}_i^{j\star}\rangle\Big),
$$

W. W. Ananduta, *Non-centralized Optimization-Based Control Schemes for Large-Scale Energy Systems*, Springer Theses,
https://doi.org/10.1007/978-3-030-89803-8

is equivalent to

$$\begin{aligned}
&2\sum_{\{i,j\}\in\mathcal{E}}\Big(\langle\hat{\boldsymbol{v}}_j^i(k)-\boldsymbol{v}_j^i(k)+\hat{\boldsymbol{v}}_i^j(k)-\boldsymbol{v}_i^j(k),\,\hat{\boldsymbol{v}}_i^j(k)+\hat{\boldsymbol{v}}_i^{j\star}\rangle\\
&+\langle\hat{\boldsymbol{v}}_j^i(k)-\boldsymbol{v}_j^i(k)+\hat{\boldsymbol{v}}_i^j(k)-\boldsymbol{v}_i^j(k),\,\hat{\boldsymbol{v}}_j^i(k)+\hat{\boldsymbol{v}}_j^{i\star}\rangle\Big)\\
&=\sum_{\{i,j\}\in\mathcal{E}}\langle\hat{\boldsymbol{v}}_j^i(k)-\boldsymbol{v}_j^i(k)+\hat{\boldsymbol{v}}_i^j(k)-\boldsymbol{v}_i^j(k),\,\hat{\boldsymbol{v}}_i^j(k)+\hat{\boldsymbol{v}}_j^i(k)\rangle\\
&=\sum_{i\in\mathcal{N}}\sum_{j\in\mathcal{N}_i}\langle\hat{\boldsymbol{v}}_i^j(k)+\hat{\boldsymbol{v}}_j^i(k)-\boldsymbol{v}_i^j(k)-\boldsymbol{v}_j^i(k),\,\hat{\boldsymbol{v}}_i^j(k)+\hat{\boldsymbol{v}}_j^i(k)\rangle,
\end{aligned}$$

where the first equality is obtained since $\boldsymbol{v}_i^{j\star}+\boldsymbol{v}_j^{i\star}=0$. Thus, by applying the two preceding manipulations to (A.1), it follows that

$$\begin{aligned}
&\sum_{i\in\mathcal{N}}\sum_{j\in\mathcal{N}_i}\Big(\langle\boldsymbol{\lambda}_i^j(k)-\boldsymbol{\lambda}_i^{j\star},\,\hat{\boldsymbol{v}}_i^j(k)+\hat{\boldsymbol{v}}_j^i(k)\rangle+2\langle\hat{\boldsymbol{v}}_i^j(k)-\boldsymbol{v}_i^j(k),\,\boldsymbol{v}_i^j(k)-\boldsymbol{v}_i^{j\star}\rangle\Big)\\
&\le\sum_{i\in\mathcal{N}}\Big(-\sigma_i\|\boldsymbol{u}_i(k+1)-\boldsymbol{u}_i^\star\|_2^2+\sum_{j\in\mathcal{N}_i}\langle\hat{\boldsymbol{v}}_i^j(k)+\hat{\boldsymbol{v}}_j^i(k)-\boldsymbol{v}_i^j(k)-\boldsymbol{v}_j^i(k),\,\hat{\boldsymbol{v}}_i^j(k)+\hat{\boldsymbol{v}}_j^i(k)\rangle\\
&\quad-\sum_{j\in\mathcal{N}_i}\Big(\|\hat{\boldsymbol{v}}_i^j(k)+\hat{\boldsymbol{v}}_j^i(k)\|_2^2+2\|\hat{\boldsymbol{v}}_i^j(k)-\boldsymbol{v}_i^j(k)\|_2^2\Big)\Big).
\end{aligned}\tag{A.2}$$

Furthermore, adding the term $\sum_{i\in\mathcal{N}}\sum_{j\in\mathcal{N}_i}(1-\eta_i^j)\langle\boldsymbol{v}_i^j(k)+\boldsymbol{v}_j^i(k),\,\hat{\boldsymbol{v}}_i^j(k)+\hat{\boldsymbol{v}}_j^i(k)\rangle$ to both sides of the inequality in (A.2) and recalling the definition of $\tilde{\boldsymbol{\lambda}}_i^j(k)$ in (3.17), it follows that

$$\begin{aligned}
&\sum_{i\in\mathcal{N}}\sum_{j\in\mathcal{N}_i}\langle\tilde{\boldsymbol{\lambda}}_i^j(k)-\boldsymbol{\lambda}_i^{j\star},\,\hat{\boldsymbol{v}}_i^j(k)+\hat{\boldsymbol{v}}_j^i(k)\rangle\\
&\quad+2\sum_{i\in\mathcal{N}}\sum_{j\in\mathcal{N}_i}\langle\hat{\boldsymbol{v}}_i^j(k)-\boldsymbol{v}_i^j(k),\,\boldsymbol{v}_i^j(k)-\boldsymbol{v}_i^{j\star}\rangle\\
&\le\sum_{i\in\mathcal{N}}\Big(-\sigma_i\|\boldsymbol{u}_i(k+1)-\boldsymbol{u}_i^\star\|_2^2\\
&\quad-\sum_{j\in\mathcal{N}_i}\Big(\|\hat{\boldsymbol{v}}_i^j(k)+\hat{\boldsymbol{v}}_j^i(k)\|_2^2+2\|\hat{\boldsymbol{v}}_i^j(k)-\boldsymbol{v}_i^j(k)\|_2^2\Big)\\
&\quad+\sum_{j\in\mathcal{N}_i}\langle\hat{\boldsymbol{v}}_i^j(k)+\hat{\boldsymbol{v}}_j^i(k)-\boldsymbol{v}_i^j(k)-\boldsymbol{v}_j^i(k),\,\hat{\boldsymbol{v}}_i^j(k)+\hat{\boldsymbol{v}}_j^i(k)\rangle\\
&\quad+\sum_{j\in\mathcal{N}_i}(1-\eta_i^j)\langle\boldsymbol{v}_i^j(k)+\boldsymbol{v}_j^i(k),\,\hat{\boldsymbol{v}}_i^j(k)+\hat{\boldsymbol{v}}_j^i(k)\rangle\Big).
\end{aligned}\tag{A.3}$$

Now, consider the last two terms on the right-hand side of the inequality in (A.3). By adding them with $\sum_{i\in\mathcal{N}}\sum_{j\in\mathcal{N}_i}(1-\eta_i^j)\left(\|\hat{\boldsymbol{v}}_i^j(k)+\hat{\boldsymbol{v}}_j^i(k)\|_2^2-\|\hat{\boldsymbol{v}}_i^j(k)+\hat{\boldsymbol{v}}_j^i(k)\|_2^2\right)=0$, it is obtain that

$$\begin{aligned}
&\sum_{i\in\mathcal{N}}\sum_{j\in\mathcal{N}_i}\Big((1-\eta_i^j)\langle \boldsymbol{v}_i^j(k)+\boldsymbol{v}_j^i(k),\hat{\boldsymbol{v}}_i^j(k)+\hat{\boldsymbol{v}}_j^i(k)\rangle\\
&\quad+\langle \hat{\boldsymbol{v}}_i^j(k)+\hat{\boldsymbol{v}}_j^i(k)-\boldsymbol{v}_i^j(k)-\boldsymbol{v}_j^i(k),\hat{\boldsymbol{v}}_i^j(k)+\hat{\boldsymbol{v}}_j^i(k)\rangle\Big)\\
&=\sum_{i\in\mathcal{N}}\sum_{j\in\mathcal{N}_i}\Big((1-\eta_i^j)\langle \boldsymbol{v}_i^j(k)+\boldsymbol{v}_j^i(k),\hat{\boldsymbol{v}}_i^j(k)+\hat{\boldsymbol{v}}_j^i(k)\rangle\\
&\quad+\langle \hat{\boldsymbol{v}}_i^j(k)+\hat{\boldsymbol{v}}_j^i(k)-\boldsymbol{v}_i^j(k)-\boldsymbol{v}_j^i(k),\hat{\boldsymbol{v}}_i^j(k)+\hat{\boldsymbol{v}}_j^i(k)\rangle(1-\eta_i^j+\eta_i^j)\Big)\\
&=\sum_{i\in\mathcal{N}}\sum_{j\in\mathcal{N}_i}\Big((1-\eta_i^j)\|\hat{\boldsymbol{v}}_i^j(k)+\hat{\boldsymbol{v}}_j^i(k)\|_2^2+\\
&\quad\eta_i^j\langle \hat{\boldsymbol{v}}_i^j(k)+\hat{\boldsymbol{v}}_j^i(k)-\boldsymbol{v}_i^j(k)-\boldsymbol{v}_j^i(k),\hat{\boldsymbol{v}}_i^j(k)+\hat{\boldsymbol{v}}_j^i(k)\rangle\Big).
\end{aligned}$$

Therefore, (A.3) becomes

$$\begin{aligned}
&\sum_{i\in\mathcal{N}}\sum_{j\in\mathcal{N}_i}\Big(\langle\tilde{\boldsymbol{\lambda}}_i^j(k)-\boldsymbol{\lambda}_i^{j\star},\hat{\boldsymbol{v}}_i^j(k)+\hat{\boldsymbol{v}}_j^i(k)\rangle+2\langle\hat{\boldsymbol{v}}_i^j(k)-\boldsymbol{v}_i^j(k),\boldsymbol{v}_i^j(k)-\boldsymbol{v}_i^{j\star}\rangle\Big)\\
&\le\sum_{i\in\mathcal{N}}\Big(-\sigma_i\|\boldsymbol{u}_i(k+1)-\boldsymbol{u}_i^\star\|_2^2-\sum_{j\in\mathcal{N}_i}\Big(\eta_i^j\|\hat{\boldsymbol{v}}_i^j(k)+\hat{\boldsymbol{v}}_j^i(k)\|_2^2+2\|\boldsymbol{v}_i^j(k)-\hat{\boldsymbol{v}}_i^j(k)\|_2^2\Big)\\
&\quad+\sum_{j\in\mathcal{N}_i}\eta_i^j\langle \hat{\boldsymbol{v}}_i^j(k)+\hat{\boldsymbol{v}}_j^i(k)-\boldsymbol{v}_i^j(k)-\boldsymbol{v}_j^i(k),\hat{\boldsymbol{v}}_i^j(k)+\hat{\boldsymbol{v}}_j^i(k)\rangle\Big).
\end{aligned}\tag{A.4}$$

Now, compute an upper-bound of the term

$$\sum_{i\in\mathcal{N}}\sum_{j\in\mathcal{N}_i}\eta_i^j\langle \hat{\boldsymbol{v}}_i^j(k)+\hat{\boldsymbol{v}}_j^i(k)-\boldsymbol{v}_i^j(k)-\boldsymbol{v}_j^i(k),\hat{\boldsymbol{v}}_i^j(k)+\hat{\boldsymbol{v}}_j^i(k)\rangle,$$

on the right-hand side of the inequality in (A.4). To that end, this term can be written as

$$\sum_{i\in\mathcal{N}}\sum_{j\in\mathcal{N}_i}-(\eta_i^j+\eta_j^i)\langle \boldsymbol{v}_i^j(k)-\hat{\boldsymbol{v}}_i^j(k),\hat{\boldsymbol{v}}_i^j(k)+\hat{\boldsymbol{v}}_j^i(k)\rangle,$$

since

$$\begin{aligned}
&\sum_{i\in\mathcal{N}}\sum_{j\in\mathcal{N}_i}\eta_i^j\langle\hat{\boldsymbol{v}}_i^j(k)+\hat{\boldsymbol{v}}_j^i(k)-\boldsymbol{v}_i^j(k)-\boldsymbol{v}_j^i(k),\hat{\boldsymbol{v}}_i^j(k)+\hat{\boldsymbol{v}}_j^i(k)\rangle\\
&=\sum_{\{i,j\}\in\mathcal{E}}\Big(\eta_i^j\langle\hat{\boldsymbol{v}}_i^j(k)+\hat{\boldsymbol{v}}_j^i(k)-\boldsymbol{v}_i^j(k)-\boldsymbol{v}_j^i(k),\hat{\boldsymbol{v}}_i^j(k)+\hat{\boldsymbol{v}}_j^i(k)\rangle\\
&\qquad+\eta_j^i\langle\hat{\boldsymbol{v}}_i^j(k)+\hat{\boldsymbol{v}}_j^i(k)-\boldsymbol{v}_i^j(k)-\boldsymbol{v}_j^i(k),\hat{\boldsymbol{v}}_i^j(k)+\hat{\boldsymbol{v}}_j^i(k)\rangle\Big)\\
&=\sum_{i\in\mathcal{N}}\sum_{j\in\mathcal{N}_i}-(\eta_i^j+\eta_j^i)\langle\boldsymbol{v}_i^j(k)-\hat{\boldsymbol{v}}_i^j(k),\hat{\boldsymbol{v}}_i^j(k)+\hat{\boldsymbol{v}}_j^i(k)\rangle.
\end{aligned}$$

Using the fact that, for any $\xi\in\mathbb{R}$, $a\in\mathbb{R}^n$, $b\in\mathbb{R}^n$, $\|a+\xi b\|_2^2=\|a\|_2^2+\xi^2\|b\|_2^2+2\xi\langle a,b\rangle \implies -\xi\langle a,b\rangle\le\frac{1}{2}\left(\|a\|_2^2+\xi^2\|b\|_2^2\right)$, an upper-bound of the term inside the summation is obtained, as follows:

$$\begin{aligned}
&-(\eta_i^j+\eta_j^i)\langle\boldsymbol{v}_i^j(k)-\hat{\boldsymbol{v}}_i^j(k),\hat{\boldsymbol{v}}_i^j(k)+\hat{\boldsymbol{v}}_j^i(k)\rangle\\
&\le\frac{1}{2}\left(\|\boldsymbol{v}_i^j(k)-\hat{\boldsymbol{v}}_i^j(k)\|_2^2+(\eta_i^j+\eta_j^i)^2\|\hat{\boldsymbol{v}}_i^j(k)+\hat{\boldsymbol{v}}_j^i(k)\|_2^2\right).
\end{aligned}$$

Therefore, using the above upper-bound and the fact that

$$\sum_{i\in\mathcal{N}}\sum_{j\in\mathcal{N}_i}\eta_i^j\|\hat{\boldsymbol{v}}_i^j(k)+\hat{\boldsymbol{v}}_j^i(k)\|_2^2=\sum_{i\in\mathcal{N}}\sum_{j\in\mathcal{N}_i}\frac{\eta_i^j+\eta_j^i}{2}\|\hat{\boldsymbol{v}}_i^j(k)+\hat{\boldsymbol{v}}_j^i(k)\|_2^2,$$

the desired inequality (3.18) follows. □

A.2 Proof of Lemma 5.2

The sequence $\{\tilde{V}_2(k)\}$ is non-negative since $\tilde{V}_2(k)$ is a sum of norms and η_i^j and α_{ij} are positive. Then, denote by $\mathcal{F}_2(k)$ the filtration up to and including the iteration k, i.e., $\mathcal{F}_2(k)=\{\mathcal{A}(\ell),\mathcal{E}^c(\ell),\boldsymbol{u}(\ell),\boldsymbol{v}(\ell),\boldsymbol{\mu}(\ell),\boldsymbol{y}(\ell),\ \ell=0,1,\ldots,k\}$. Now, it must be shown that the conditional expectation of the sequence with respect to $\mathcal{F}_2(k)$ is always non-increasing. Based on Assumptions 5.1 and 5.2, the initialization step in Algorithm 5.2, and the update rules (5.7) and (5.8), the variables $\boldsymbol{y}_i^j(k+1)$ and $\boldsymbol{\mu}_i^j(k+1)$, for each $j\in\mathcal{N}_i\cap\mathcal{M}_p$ and $i\in\mathcal{M}_p$, are updated when agent $\phi(i,\cdot)$ is active, whereas, for each $j\in\mathcal{N}_i\backslash\mathcal{M}_p$ and $i\in\mathcal{M}_p$, those variables are only updated when agents $\phi(i,\cdot)$ and $\phi(j,\cdot)$ are active and communication link $\{\phi(i,\cdot),\phi(j,\cdot)\}$ is active. Therefore, it is observed from (5.10) that the probability of $\boldsymbol{y}_i^j(k+1)$ and $\boldsymbol{\mu}_i^j(k+1)$ being updated is $\alpha_{ij}\in(0,1]$, whereas, with probability $1-\alpha_{ij}$, they are not updated and the values remain the same as $\boldsymbol{y}_i^j(k)$ and $\boldsymbol{\mu}_i^j(k)$. Hence, it is obtained with probability 1 that

$$\begin{aligned}
&\mathbb{E}\left(\tilde{V}_2(k+1)|\mathcal{F}_1(k)\right)-\tilde{V}_2(k)\\
&=\mathbb{E}\left(\frac{1}{2}\|\boldsymbol{\mu}(k+1)-\boldsymbol{\mu}^\star\|_{\tilde{H}}^2+\frac{1}{2}\|\boldsymbol{y}(k+1)-\boldsymbol{v}^\star\|_{\tilde{G}}^2\middle|\mathcal{F}_1(k)\right)\\
&\quad-\tilde{V}_2(k)\\
&=\sum_{i\in\mathcal{N}}\sum_{j\in\mathcal{N}_i}\left(\frac{\alpha_{ij}}{2\alpha_{ij}\eta_i^j}\|\boldsymbol{\mu}_i^j(k+1)-\boldsymbol{\mu}_i^{j\star}\|_2^2+\frac{1-\alpha_{ij}}{2\alpha_{ij}\eta_i^j}\|\boldsymbol{\mu}_i^j(k)-\boldsymbol{\mu}_i^{j\star}\|_2^2\right.\\
&\quad\left.+\frac{\alpha_{ij}}{2\alpha_{ij}}\|\boldsymbol{y}_i^j(k+1)-\boldsymbol{y}_i^{j\star}\|_2^2+\frac{1-\alpha_{ij}}{2\alpha_{ij}}\|\boldsymbol{y}_i^j(k)-\boldsymbol{v}_i^{j\star}\|_2^2\right)-\tilde{V}_2(k)\\
&=\frac{1}{2}\|\boldsymbol{\mu}(k+1)-\boldsymbol{\mu}^\star\|_H^2-\frac{1}{2}\|\boldsymbol{\mu}(k)-\boldsymbol{\mu}^\star\|_H^2\\
&\quad+\frac{1}{2}\|\boldsymbol{y}(k+1)-\boldsymbol{y}^\star\|_2^2-\frac{1}{2}\|\boldsymbol{y}(k)-\boldsymbol{y}^\star\|_2^2,\\
&=V_2(k+1)-V_2(k),
\end{aligned}$$

where the last equality is obtained since the terms $\sum_{i\in\mathcal{N}}\sum_{j\in\mathcal{N}_i}\frac{1}{2\alpha_{ij}}\|\boldsymbol{y}_i^j(k)-\boldsymbol{y}_i^{j\star}\|_2^2$, $\sum_{i\in\mathcal{N}}\sum_{j\in\mathcal{N}_i}\frac{1}{2\alpha_{ij}\eta_i^j}\|\boldsymbol{\mu}_i^j(k)-\boldsymbol{\mu}_i^{j\star}\|_2^2$, and $-\tilde{V}_2(k)$ cancel each other. Therefore, by applying (3.52) to this relation, the desired relations in (5.11) follows, with probability 1, when $\eta_i^j =\in (0,1)$. Thus, (5.21) also shows that the sequence $\{\tilde{V}_2(k)\}$ is non-negative supermartingale. □

A.3 Proof of Theorem 5.3

By rearranging and iterating the first inequality in (5.21) over $\ell = 0, 1, \ldots, k$ and taking the total expectation, it follows that

$$\begin{aligned}
&\sum_{\ell=0}^{k}\left(\frac{1}{2}\mathbb{E}\left(\|\boldsymbol{r}(\ell+1)+\boldsymbol{y}(\ell+1)-\boldsymbol{y}(\ell)\|_2^2\right)+\sum_{i\in\mathcal{N}}\sigma_i\mathbb{E}\left(\|\boldsymbol{u}_i(\ell+1)-\boldsymbol{u}_i^\star\|_2^2\right)\right.\\
&\qquad\left.+\sum_{i\in\mathcal{N}}\sum_{j\in\mathcal{N}_i}\frac{1-\eta_i^j}{2}\mathbb{E}\left(\|\boldsymbol{r}_i^j(\ell+1)\|_2^2\right)\right)\leq\sum_{\ell=0}^{k}\mathbb{E}\left(\left(\tilde{V}_2(\ell)-\tilde{V}_2(\ell+1)\right)\right)\\
&=\tilde{V}_2(0)-\mathbb{E}(\tilde{V}_2(k+1))\leq\tilde{V}_2(0),
\end{aligned}\tag{A.5}$$

where the last inequality is obtained by dropping the non-positive term $\mathbb{E}(\tilde{V}_2(k+1))$. The inequalities in (A.5) implies that the terms on the left-hand side of the first inequality are summable and converge to 0 as k goes to infinity i.e.,

$$\lim_{k\to\infty}\left(\frac{1}{2}\mathbb{E}\left(\|\boldsymbol{r}(\ell+1)+\boldsymbol{y}(\ell+1)-\boldsymbol{y}(\ell)\|_2^2\right)+\sum_{i\in\mathcal{N}}\sigma_i\mathbb{E}\left(\|\boldsymbol{u}_i(\ell+1)-\boldsymbol{u}_i^\star\|_2^2\right)\right.$$
$$\left.+\sum_{i\in\mathcal{N}}\sum_{j\in\mathcal{N}_i}\frac{1-\eta_i^j}{2}\mathbb{E}\left(\|\boldsymbol{r}_i^j(\ell+1)\|_2^2\right)\right)=0.$$

Therefore, based on the preceding relation and the Markov inequality, it holds that

$$\limsup_{k\to\infty}\mathbb{P}\left(\frac{1}{2}\|\boldsymbol{r}(\ell+1)+\boldsymbol{y}(\ell+1)-\boldsymbol{y}(\ell)\|_2^2+\sum_{i\in\mathcal{N}}\sigma_i\|\boldsymbol{u}_i(\ell+1)-\boldsymbol{u}_i^\star\|_2^2\right.$$
$$\left.+\sum_{i\in\mathcal{N}}\sum_{j\in\mathcal{N}_i}\frac{1-\eta_i^j}{2}\|\boldsymbol{r}_i^j(\ell+1)\|_2^2\geq\varepsilon\right)$$
$$\leq\limsup_{k\to\infty}\frac{1}{\varepsilon}\mathbb{E}\left(\frac{1}{2}\|\boldsymbol{r}(\ell+1)+\boldsymbol{y}(\ell+1)-\boldsymbol{y}(\ell)\|_2^2+\sum_{i\in\mathcal{N}}\sigma_i\|\boldsymbol{u}_i(\ell+1)-\boldsymbol{u}_i^\star\|_2^2\right.$$
$$\left.+\sum_{i\in\mathcal{N}}\sum_{j\in\mathcal{N}_i}\frac{1-\eta_i^j}{2}\|\boldsymbol{r}_i^j(\ell+1)\|_2^2\right)=0,$$

implying that, the convergence is with probability 1, i.e.,

$$\lim_{k\to\infty}\|\boldsymbol{r}(k+1)+\boldsymbol{y}(k+1)-\boldsymbol{y}(k)\|_2^2=0, \tag{A.6}$$
$$\lim_{k\to\infty}\sigma_i\|\boldsymbol{u}_i(k+1)-\boldsymbol{u}_i^\star\|_2^2=0,\quad \forall i\in\mathcal{N}, \tag{A.7}$$
$$\lim_{k\to\infty}\|\boldsymbol{r}(k+1)\|_2^2=\lim_{k\to\infty}\|\boldsymbol{v}(k+1)-\boldsymbol{y}(k+1)\|_2^2=0, \tag{A.8}$$

with probability 1. Moreover, by (A.6) and (A.8), it holds that

$$\lim_{k\to\infty}\|\boldsymbol{y}(k+1)-\boldsymbol{y}(k)\|_2^2=0, \tag{A.9}$$

with probability 1. Additionally, due to the initialization step in Algorithm 5.2 and the update rule in (5.7), $\boldsymbol{y}_i^j(k)=-\boldsymbol{y}_j^i(k)$, for all $j\in\mathcal{N}_i$, $i\in\mathcal{N}$, and $k\in\mathbb{Z}_{\geq 0}$. Based on this fact and the relation in (A.8), it also holds with probability 1 that

$$\lim_{k\to\infty}\|\boldsymbol{v}_i^j(k+1)+\boldsymbol{v}_j^i(k+1)\|_2^2=0,\ \forall j\in\mathcal{N}_i,\ \forall i\in\mathcal{N}. \tag{A.10}$$

Furthermore, based on (5.21), the sequences $\{\boldsymbol{y}(k)\}$ and $\{\boldsymbol{\mu}(k))\}$ are bounded with probability 1 and have accumulation points. Additionally, based on the boundedness of $\{\boldsymbol{y}(k)\}$ and (A.8), $\{\boldsymbol{v}(k)\}$ is also bounded with probability 1 and has accumulation points, which are equal to that of $\{\boldsymbol{y}(k)\}$. Now, consider that the subsequence $\{(\boldsymbol{v}(k_\ell),\boldsymbol{y}(k_\ell),\boldsymbol{\mu}(k_\ell))\}$ converges to the limit point denoted by $(\boldsymbol{v}^{\mathrm{a}},\boldsymbol{y}^{\mathrm{a}},\boldsymbol{\mu}^{\mathrm{a}})$, where $\boldsymbol{v}^{\mathrm{a}}=\boldsymbol{y}^{\mathrm{a}}$. Then, it will be shown that the point $(\boldsymbol{u}^\star,\boldsymbol{v}^{\mathrm{a}},\boldsymbol{y}^{\mathrm{a}},\boldsymbol{\mu}^{\mathrm{a}})$ is a saddle point of the

Lagrangian function $L_2(\boldsymbol{u}, \boldsymbol{v}, \boldsymbol{y}, \boldsymbol{\mu})$ as defined in (3.29), i.e., $(\boldsymbol{u}^\star, \boldsymbol{v}^{\mathrm{a}}, \boldsymbol{y}^{\mathrm{a}}, \boldsymbol{\mu}^{\mathrm{a}})$ satisfies the saddle point inequalities in (3.31).

Based on (A.8), it holds that $\boldsymbol{v}_i^{j\mathrm{a}} - \boldsymbol{y}_i^{j\mathrm{a}} = \lim_{\ell \to \infty}(\boldsymbol{v}_i^j(k_\ell) - \boldsymbol{y}_i^j(k_\ell)) = 0$, with probability 1, for all $j \in \mathcal{N}_i$ and $i \in \mathcal{N}$. Therefore, for any $\boldsymbol{\mu} \in \mathbb{R}^{\sum_{i\in\mathcal{N}} h|\mathcal{N}_i|}$, $L_2(\boldsymbol{u}^\star, \boldsymbol{v}^{\mathrm{a}}, \boldsymbol{y}^{\mathrm{a}}, \boldsymbol{\mu}) = L_2(\boldsymbol{u}^\star, \boldsymbol{v}^{\mathrm{a}}, \boldsymbol{y}^{\mathrm{a}}, \boldsymbol{\mu}^{\mathrm{a}})$, satisfying the first inequality in (3.31). Now, by substituting k with k_ℓ in (5.6), for all $i \in \mathcal{N}$, and taking the limit as ℓ goes to infinity, it holds with probability 1 that

$$
\begin{aligned}
(\boldsymbol{u}^\star, \boldsymbol{v}^{\mathrm{a}}) &= \lim_{\ell\to\infty} \arg \min_{(\boldsymbol{u}_i,\boldsymbol{v}_i)\in\mathcal{L}_i, i\in\mathcal{N}} \sum_{i\in\mathcal{N}} \Big(f_i^{\mathrm{l}}(\boldsymbol{u}_i) + f_i^{\mathrm{c}}(\boldsymbol{v}_i) \\
&\quad + \sum_{j\in\mathcal{N}_i} \Big(\langle \boldsymbol{\mu}_i^j(k_\ell), \boldsymbol{v}_i^j \rangle + \|\boldsymbol{v}_i^j - \boldsymbol{y}_i^j(k_\ell)\|_2^2 \Big)\Big) \\
&= \arg \min_{(\boldsymbol{u}_i,\boldsymbol{v}_i)\in\mathcal{L}_i, i\in\mathcal{N}} \sum_{i\in\mathcal{N}} \Big(f_i^{\mathrm{l}}(\boldsymbol{u}_i) + f_i^{\mathrm{c}}(\boldsymbol{v}_i) + \sum_{j\in\mathcal{N}_i} \Big(\langle \boldsymbol{\mu}_i^{j\mathrm{a}}, \boldsymbol{v}_i^j \rangle + \|\boldsymbol{v}_i^j - \boldsymbol{y}_i^{j\mathrm{a}}\|_2^2 \Big)\Big) \\
&= \arg \min_{(\boldsymbol{u}_i,\boldsymbol{v}_i)\in\mathcal{L}_i, i\in\mathcal{N}} \sum_{i\in\mathcal{N}} \Big(f_i^{\mathrm{l}}(\boldsymbol{u}_i) + f_i^{\mathrm{c}}(\boldsymbol{v}_i) + \sum_{j\in\mathcal{N}_i} \langle \boldsymbol{\mu}_i^{j\mathrm{a}}, \boldsymbol{v}_i^j \rangle - \langle \boldsymbol{\mu}_i^{j\mathrm{a}}, \boldsymbol{y}_i^{j\mathrm{a}} \rangle \Big),
\end{aligned}
\tag{A.11}
$$

where the second equality is obtained by using (A.9) and the last equality is obtained by introducing the constant $-\sum_{i\in\mathcal{N}}\sum_{j\in\mathcal{N}_i}\langle \boldsymbol{\mu}_i^{j\mathrm{a}}, \boldsymbol{y}_i^{j\mathrm{a}}\rangle$ and since the quadratic term $\sum_{i\in\mathcal{N}}\sum_{j\in\mathcal{N}_i}\|\boldsymbol{v}_i^j - \boldsymbol{y}_i^{j\mathrm{a}}\|_2^2$ is 0 at $(\boldsymbol{u}^\star, \boldsymbol{v}^{\mathrm{a}})$ due to (A.8). Now, note that $(\boldsymbol{v}^{\mathrm{a}}, \boldsymbol{y}^{\mathrm{a}})$ minimizes $\sum_{i\in\mathcal{N}}\sum_{j\in\mathcal{N}_i}\|\boldsymbol{v}_i^j - \boldsymbol{y}_i^j\|_2^2$ since $\boldsymbol{v}_i^{j\mathrm{a}} - \boldsymbol{y}_i^{j\mathrm{a}} = 0$, for all $j \in \mathcal{N}_i$ and $i \in \mathcal{N}$. Therefore, $(\boldsymbol{u}^\star, \boldsymbol{v}^{\mathrm{a}}, \boldsymbol{y}^{\mathrm{a}})$ minimizes $L_2(\boldsymbol{u}, \boldsymbol{v}, \boldsymbol{y}, \boldsymbol{\mu}^{\mathrm{a}})$, implying the satisfaction of the second inequality in (3.31). Hence, it can be concluded that $(\boldsymbol{u}^\star, \boldsymbol{v}^{\mathrm{a}}, \boldsymbol{y}^{\mathrm{a}}, \boldsymbol{\mu}^{\mathrm{a}})$ is a saddle point of $L_1(\boldsymbol{u}, \boldsymbol{v}, \boldsymbol{y}, \boldsymbol{\mu})$. Finally, set $\boldsymbol{y}^\star = \boldsymbol{y}^{\mathrm{a}}$ and $\boldsymbol{\mu}^\star = \boldsymbol{\mu}^{\mathrm{a}}$ in $\tilde{V}_2(k)$ (see (5.20)). Since the subsequence of $\tilde{V}_2(k_\ell)$ converges to 0 and $\tilde{V}_2(k)$ is non-negative supermartingale, the entire sequence $\{(\boldsymbol{u}(k), \boldsymbol{y}(k), \boldsymbol{\mu}(k))\}$ converges to $(\boldsymbol{v}^{\mathrm{a}}, \boldsymbol{y}^{\mathrm{a}}, \boldsymbol{\mu}^{\mathrm{a}})$ with probability 1. □

A.4 Proof of Theorem 5.4

Recall that rearranging the summation of (3.52) over $\ell = 0, 1, \ldots, k-1$ and taking the total expectation yields

$$
\begin{aligned}
&\sum_{\ell=0}^{k-1} \Big(\frac{1}{2}\mathbb{E}\left(\|\boldsymbol{v}(\ell+1) - \boldsymbol{y}(\ell)\|_2^2\right) + \sum_{i\in\mathcal{N}} \sigma_i \mathbb{E}\left(\|\boldsymbol{u}_i(\ell+1) - \boldsymbol{u}_i^\star\|_2^2\right) \\
&\quad + \sum_{i\in\mathcal{N}}\sum_{j\in\mathcal{N}_i} \frac{1-\eta_i^j}{2}\mathbb{E}\left(\|\boldsymbol{v}(\ell+1) - \boldsymbol{y}(\ell+1)\|_2^2\right) \Big)
\end{aligned}
$$

$$\leq \tilde{V}_2(0) = \frac{1}{2}\|\boldsymbol{y}(0) - \boldsymbol{y}^\star\|_G^2 + \frac{1}{2}\|\boldsymbol{\mu}(0) - \boldsymbol{\mu}^\star\|_{\tilde{H}}^2$$
$$\leq \frac{1}{\underline{\alpha}}\left(\frac{1}{2}\|\boldsymbol{y}(0) - \boldsymbol{y}^\star\|_2^2 + \frac{1}{2}\|\boldsymbol{\mu}(0) - \boldsymbol{\mu}^\star\|_H^2\right), \quad \text{(A.12)}$$

where the term $\boldsymbol{r}(k)$ is substituted by $\boldsymbol{v}(k) - \boldsymbol{y}(k)$ using (3.39) and the last inequality is obtained by using the definition of $\underline{\alpha}$. Moreover, by the convexity of the squared of the Euclidean norm, it holds that

$$k\mathbb{E}\left(\|\bar{\boldsymbol{v}}(k) - \bar{\boldsymbol{y}}(k-1)\|_2^2\right) \leq \sum_{\ell=0}^{k-1}\mathbb{E}\left(\|\boldsymbol{v}(\ell+1) - \boldsymbol{y}(\ell)\|_2^2\right),$$
$$k\mathbb{E}\left(\|\bar{\boldsymbol{u}}_i(k) - \boldsymbol{u}_i^\star\|_2^2\right) \leq \sum_{\ell=0}^{k-1}E\left(\|\boldsymbol{u}_i(\ell+1) - \boldsymbol{u}_i^\star\|_2^2\right),$$
$$k\mathbb{E}\left(\|\bar{\boldsymbol{v}}(k) - \bar{\boldsymbol{y}}(k)\|_2^2\right) \leq \sum_{\ell=0}^{k-1}\mathbb{E}\left(\|\boldsymbol{v}(\ell+1) - \boldsymbol{y}(\ell+1)\|_2^2\right).$$

By applying the three preceding relations to (A.12), the desired inequality (5.28) follows since $\eta_{ij} \in (0, 1)$, for all $j \in \mathcal{N}_i$ and $i \in \mathcal{N}$. □

Author Biography

W. Wicak Ananduta is currently a postdoctoral researcher at Delft Center for Systems and Control (DCSC), TU Delft, in the Netherlands. This book is the result of his Ph.D. thesis in automatic control, robotics and vision, which he successfully defended in December 2019 and thus, earned him a Ph.D. degree from Universitat Politècnica de Catalunya in Barcelona, Spain. During his Ph.D. training, he was fully supported by a Marie Skłodowska-Curie innovative training network from the European Union. In 2018, he visited Tokyo Institute of Technology to work with Professor Hideaki Ishii and Professor José Maria Maestre for three months. He also visited Arizona State University to work with Professor Angelia Nedić in 2019. Additionally, he obtained a master degree in systems and control from TU Delft in 2016 and a bachelor degree in electrical engineering from University of Indonesia in 2011. His research interests include distributed optimization, equilibrium seeking algorithms, and their applications in large-scale systems.